Technology Education, Innovation, and Management

Springer
*Berlin
Heidelberg
New York
Barcelona
Budapest
Hong Kong
London
Milan
Paris
Santa Clara
Singapore
Tokyo*

K. Langer M. Metzing D. Wahl (Eds.)

Technology Education, Innovation, and Management

Proceedings of the WOCATE Conference 1994

With 53 Figures and 18 Tables

Springer

Dr. Kati Langer
Dr. Matthias Metzing
Dr. Detlef Wahl

WOCATE
Schlösserstraße 9
D-99084 Erfurt

```
Library of Congress Cataloging-in-Publication Data
Technology education, innovation, and management : proceedings of
  WOCATE conference / [edited by] K. Langer, Metzing, M., D. Wahl.
      p.   cm.
    "Sponsored by the World Council of Associations for Technology
  Education (WOCATE)"--T.p. verso.
    Conference held from Sept. 24 to 29, 1994 in Banská Bystrica,
  Slovak Republic.
    Includes bibliographical references and index.
    ISBN 3-540-60018-3 (Berlin : soft : acid free-paper)
    1. Technology--Study and teaching--Congresses.  2. Technological
  innovations--Management--Congresses.   I. Langer, K. (Katrin), 1965-
  .  II. Metzing, M. (Matthias), 1964-   .  III. Wahl, D. (Detlef),
  1963-    .  IV. World Council of Associations for Technology
  Education.
    T62.T45 1995
    607.1--dc20                                              95-30836
                                                                  CIP
```

Sponsored by the World Council of Associations
for Technology Education (WOCATE)

CR Subject Classification (1991): K.3, H.4, J.6

ISBN 3-540-60018-3 Springer-Verlag Berlin Heidelberg New York

This work is subject to copyright. All rights are reserved, whether the whole or part of the material is concerned, specifically the rights of translation, reprinting, reuse of illustrations, recitation, broadcasting, reproduction on micro-film or in any other way, and storage in data banks. Duplication of this publication or parts thereof is permitted only under the provisions of the German Copyright Law of September 9, 1965, in its current version, and permission for use must always be obtained from Springer-Verlag. Violations are liable for prosecution under the German Copyright Law.

© Springer-Verlag Berlin Heidelberg 1995
Printed in Germany

The use of general descriptive names, trademarks, etc. in this publication does not imply, even in the absence of a specific statement, that such names are exempt from the relevant protective laws and regulations and therefore free for general use.

Cover Design: Springer-Verlag Design & Production, Heidelberg
Typesetting: Camera ready by editors
Printed on acid-free paper   SPIN 10503424   45/3142 - 5 4 3 2 1 0

Preface

In the development of the Human-Technology-Interface the connection of teaching innovation processes and value increasing is a strategic development question. Technology education is embedded in cultural and socio-economical contexts that vary from country to country. That leads to different points of view by educators and politicians.

The World Council of Associations for Technology Education (WOCATE) conference's "Technology Education, Innovation and Management" main intention was, to share the ideas of innovation management and its trends of integration in educational processes.

The conference took place from 24th to 29th September 1994 in Banská Bystrica, Slovak Republic and gave especially the colleagues from the Eastern countries the chance to present their current experiences in the process of developing technology education as an acknowledged subject. According to the topic of the conference the main focuses were:

- technology education at the policy level
- human resource development
- methods, didactic and equipment
- educational innovations
- school-industry-link-projects and
- experiences of implementing technology education into cultural contexts.

The most profound contributions of participants from 26 countries at the conference provide a concentrated overview about theoretical findings and starting points for practical work in this field. Along with a resolution towards further development of technology education the book demonstrates the existing level of international agreement.

By comparison of today's needs concepts for further work are developed. For example similarities existing in the process of teaching innovation on the one hand and the real work situation in small and medium scale industries on the other hand are discussed. This is a focus that pays attention to a general trend towards the link of value increasing and education. That idea is especially related to the discussion about shortening the years of study.

Unfortunately not all of the contributions given in three parallel paper presentations can be considered in this report. Nevertheless the interested reader should not hesitate to contact the authors of the papers listed at the end of this book or contact the WOCATE-Office in Erfurt to get further information concerning presentations that were given. We should be happy to assist you in this matter.

Acknowledgements

WOCATE would like to thank all participants, keynote-speakers, paper presenters and rapporteurs for their contributions to this conference.

It was certainly an open and warm climate, that was determined by a high standard of academic exchanges of ideas.

This climate was carried by the work of the local organisers as well as by the work of the colleagues from the WOCATE-Office in Erfurt. We also should not forget to mention the conceptional work of Prof. Dr. Dietrich Blandow, Secretary of WOCATE.

The names of Doris Traberth and Horst Wengel are representative for all colleagues who were involved in the organisation of this event.

We would like to thank the COMETT-Programme that supported the publication of this report. May this book deliver a valuable contribution to the discussion of Technology Education in the field of innovation and management. It delivers a theoretical contribution as well as starting points for practice with remarkable findings for the reader. The scientific value is completed by diagrams, figures and tables.

Finally our thanks go to the Springer Publishing House for the possibility to share the results of this conference with our colleagues.

Erfurt, June 1995
Katrin Langer
Matthias Metzing
Detlef Wahl

Table of Contents

Official Opening and Welcome

Jozef Belák, Director of the Section of Technical Policy..........................1

1. Educational Strategies for Technology Education

The development of technology education in New Zealand
Don Ferguson 6

Technology education in Germany's gymnasia
Walter E. Theuerkauf et al. 15

Perspectives and concepts in the Swedish national curriculum for technology - and a modest proposal for a general model that will explain it all - hopefully...
Thomas Ginner 32

A new curriculum for technology education in Schleswig-Holstein
Gerd Höpken 37

New education law, new curriculum
Lajos Radics 51

Research in Technology Education: some insights
George Shield 55

2. Human Resource Development for Technology Education

HRD, Innovative and Integrative Thinking of Education for Life
Dietrich Blandow and Michael J. Dyrenfurth 75

Minds 2000+: Innovations in global change as part of technology education...82
James L. Barnes

Key qualifications - contents of general and vocational education
Manfred Lutherdt 91

Strategies, methods and principles of development thinking
Bernd Hill 101

Cognitive structure of technology subjects for deciding instructional design and learning preferences
S. Swaminatha Pillai 107

Psychohygiene in a manager's work
Alfred Prigl 117

3. Methods, Didactics and Equipment for Technology Education

Developing a competency to act within networked systems
Walter E. Theuerkauf and Andreas Weiner 125

New approaches towards interdisciplinary technological enlightenment
Ingrid Lisop 137

Didactical structure of the technological culture
Vanya Georgieva 143

A teaching strategy to promote student confidence and creativity
in the design and prototyping of digital electronic systems
Clive D. Mockford 147

The ontdekplek
Diny Flierman and Harry Valkenier 155

The teaching laboratory as a meaningful environment
Matzi Eliahu 158

4. Educational Innovations for Technology Education

Education to teamwork in building technology MHPO -
method holistic participation - a contribution to technology education
Peter Schmid 162

Management and innovation
Haris Papoutsakis 169

Technological education and innovation in connection with tendencies
of a world that is changing
Ilia Natali 183

Integration as an innovation course in the education concerning
all fields of technology
Kazimierz Uzdzicki 192

Methodological initiation aspects of educational innovation
Maria Jakowicka 196

Innovation in initial teacher training: an analysis of benefits, costs and resource-related opportunities
Paul Griffiths 201

5. School-Industry-Link-Projects

Technology education, innovation and management
Alzbeta Dingová 210

Are we making technology education attractive to our students?
John Eggleston 216

Economic and industrial understanding as part of design and technology education in the primary curriculum
Clare Benson 224

Working with industry to enhance technology education
Sylvia Innes 233

School and economy - first steps in Hungary
Sándor Kiss and István Matuz 241

6. Experiences of Implementing Technology Education into Cultural Context

The objects and tasks of technology education in comprehensive schools in the Czech Republic
Frantisek Mosna 245

Development and implementation of a model for technology education in South Africa - the ORT-STEP experience
Eli Eisenberg 250

Present days problems of technology education in Slovakia
Ján Hudec and Ondrej Nemcok 259

European supporting programmes - experiences in the field of education and training from the view of East German universities and enterprises
Frank March 266

Building of a technical educational system of a non-university type in Slovakia
Pavel Prokopovic 272

What factors will influence the wider implementation of
technology education in South African Schools?
Rodney F. Sherwood 281

The introduction of technology laboratories into schools in the
disadvantaged communities of the Republic of South Africa
Arthur Cotton 291

Conference Paper Presentations: Summary 298

Resolution Adopted by the Conference 305

List of Participants 307

Subject Index 317

Official Opening and Welcome

Jozef Belák
Ministry of Economy of the Slovak Republic
Department for Technical Policy
Mierová 19, SK-82715 Bratislava, Slovak Republic

Ladies and Gentlemen,
Participants at WOCATE conference,

Allow me to bring you personal greetings from the Minister of Economy of the Slovak Republic, Mr. Magvasi, wishing you success in your negotiations.
 As a follow up to his words in a written speech delivered to you, let me inform you on the development of economy of the Slovak Republic since the WOCATE workshop was held in Poprad city in August 1993.
 In this time period it was our endeavour to proceed in the economic sphere of our Republic in the spirit of those principles which we have set forth in 1993.
Among decisive actions steps taken by the Ministry of Economy of SR, is also the effort for principal turnaround in the enterprise sphere. Especially attention was given to commercialization and ownership transformation in the enterprise sphere.
 In order to recover and unblock cash flows, we co-operate on further procedures of the financial restructuring of bad bank loans, and by that also the enterprise sphere, and that especially in the frame of EFSAL loan preparation. Another type of financial aid is a proposal to establish a joint stock company with a deposit of ECU 20 million in co-operation with the European Bank for Reconstruction and Development.
 In the area of forming a favourable business environment and promotion of private sphere development, the National Agency for Promotion of Small and Medium Business Development was established, together with advisory centres, including provision of both domestic and foreign financial resources.
 The National Agency for Promotion of Small and Medium Business Development, develops its key activities in the area of education, consulting, establishing technology and innovation centres and a subcontractor exchange.
 It was sectoral policy orientation to support restructuring and revitalization of enterprises. Essential directions of restructuring and revitalization process are characterized in the Draft Industrial Policy of SR, which was negotiated at the Economic Council of the Government of the SR on August 22, 1994.
 To facilitate privatization of armament enterprises with an adequate state participation, recovery projects were elaborated. In regard to the bad economic

situation of armament enterprises, accelerated privatization proved to be impossible. The Ministry of Economy of SR elaborated a paper "Concept of military production in the industry of SR", which was submitted for negotiation by the government of SR.

In the energy sector a proposal has been prepared and adopted for arrangement, restructuring an gradual privatization of current state enterprises with a prudent entry of foreign capital.

For completion of a large energy construction, a report on progress of construction and securing the financing of Mochovce nuclear power station, as well as proposals for joining of the Slovak Republic to the Vienna Convention on Civil Liability for Nuclear Damages were submitted for negotiations by the government of SR.

For the internal trade, principles of state policy for consumer protection were adopted. Further a methodical instruction was elaborated for supervision over the consumer protection, to be executed by territorial bodies of state administration.

With the aim to secure tourist traffic development in a way as to increase its share on creation of new jobs, as for revitalization of regions and the increase of foreign exchange revenue, measures have been adopted in the area of promotion, the quality of publishing activity was improved and new activities are being created when presenting the offer of tourist traffic on specialized exhibitions and fairs.

In the foreign trade policy an emphasis was given to a thorough reflection of conclusions from bilateral and multilateral negotiations of the Uruguay Round of GATT for economic and trade relations. A "Report on results of the Uruguay Round of multilateral business negotiations and on accession of SR to the World Trade Organization" was submitted for negotiation by the government of SR.

Multilateral and bilateral agreements were prepared and realized, necessary for expansion of business areas with 17 more countries of Europe and Asia. "Draft measures for pro-export program of SR" were elaborated to support the pro-export policy.

The development in the branches of industry of the Slovak Republic's economy in the first half of 1994, compared with last year, is slightly positive. In the industry the production rose overall by 1.9% (USD 1 million), the most in the generation and distribution of energies, gas and water. At the same time extraction of ores and coal has declined, which is a consequence of a decline program in these industries. Sale of products increased by 5.8% (USD 347 million), of which in export by 1.9% (USD 114 million). This increase in export represents its increase by 8.8%. Total export is 39.3% of the volume of products sold. 90% of that are products of processing industry.

Presently there is still a decline in the volume of civil works (by 9% compared with last year). It is a consequence of recession, slow restructuring of production capacities and restrictions on housing construction. But we assume, based on results of the first half of 1994, that the overall production of goods will increase in the second half of 1994 by 4.2%.

In the foreign trade the total turnover for the first half of 1994 increased by 17.8%. Contribution to this is an increase in export by 60%. The Slovak Republic achieved a positive balance of foreign trade with the countries of the EU, CEFTA and APEC. A negative balance was recorded with countries of EFTA and countries of the former Soviet Union.

Economic policy intentions for 1995 were adopted by the government of the SR in the past month. These intentions:

- define major problems for entering of Slovak economy into the phase of recovery. These are: fiscal deficit, absence of program of public investments and allocation mechanisms, inappropriate structure of economy for a surplus of production which is marketable, inappropriate export structure, tendency towards a negative balance of the trade balance, which is not sufficiently eliminated by increase of export services, imperfect competitive environment, broad and not transparent social network, insolvency of enterprises and increase in the number enterprises working with loss, permanent changes in the privatization concept and preference of short-term interests,

- determine goals and intentions of economic policy of the government of the SR for 1994 and 1995, and that is by getting closer to a zero growth of the Gross Domestic Product, maintaining the price growth in the interval of 10-13%, unemployment rate below 17% and not exceeding the budget deficit of Sk 14 billion (USD 437 million). In order to achieve these targets, the following development in the macro-economic indicators is expected:

i. prognosis of Gross Domestic Product for 1994 assumes moderation of its decline in an interval of 0-2.7%, as a possible consequence of starting a boom cycle and decrease in interest rates from our import trade partners (Germany, Austria, but also the Czech Republic),

ii. in the area of employment there will be no improvement, labour market will still be significantly imbalanced,

iii. it is counted on accelerated privatization (so called second wave of privatization),

iv. realization of measures for improvement of budget situation is being assumed within the interest to secure fiscal consolidation and macro-economic stability,

v. financial restructuring of enterprises and banks will take place using foreign capital, especially through EFSAL loan, for which diagnostic studies are being prepared for 20 companies,

vi. stabilization and restructuring privatization program will be realized through Joint Stock Company with the assistance from EBRD and the National Property Fund of the SR,

vii. They assume development in 1995 in a way, provided that there are unchanged conditions of development of stabilizing modifications of economic policy, the

Gross Domestic Product can reach growth of 0-1.8%, total rate of inflation 11-15%, unemployment rate 18.9-20.1%, and the deficit of the state budget should reach approx. SK 17 billion (USD 531 million).

The main goal in the intentions of economic policy for 1995 is to regain the trust in both human and economic potential of the SR and to lay foundations for a long-term sustainable economic growth and to place Slovak economy among countries of the European Union.

For this reason the priorities are as follows:

1. Program for speeding up the transformation (privatization, restructuring, privatization of banks, and development of capital market), concept for public sector development will be prepared, transformation of social sphere, health care and school system,

2. To secure macro-economic stability, especially in the area of strategic aims, as decreasing the negative trade balance and the balance of payments of the state, but also moderation of tendencies in the development of state debt and deficit of the state budget, and the co-ordination of the economic policy of the Slovak government and the National Bank of Slovakia.

3. Economic recovery, development and restructuring through industrial, agricultural, pro-export policy and a scientific, technical and technological policy, as well as the policy in the area of small and medium size businesses. There must be given space for a more emphatic realization of programs for promotion of small and medium businesses, implementation of enterprise restructuring programs, programs of energy savings and support for introductions of high-technologies, promotion programs for goods and services export growth, and support programs for production on the base of domestic raw materials and domestic production, realized through appropriate, economically effective projects in the business sphere.

Closely related to the intentions of economic policy are the ways of fulfilling them, besides others also in the area of science, technologies and training of employees. Strategic orientation in the area of science and technology has the main task:

- to facilitate transformation of the economy into a market economy by supporting inventional and innovative processes,
- to direct the care towards quality promotion of the scientific and technical potential and to make its activity more effective,
- to stimulate activities in the area of science and technology, which create presumptions for realization of economic policy intentions.

We have set forth basic principles in the strategy, from which our scientific and technological policy has to result from. These are:

- Science and technology are national priorities,

- Basic interest of the state is the development of the national inventional and innovative potential, which can be achieved by joint efforts of both scientific and technical community, economic sphere and state bodies,
- Science and technology are conditioned by each other,
- The process of transformation assumes limitation of a directive and administrative interference with the development of science and technology,
- Supportive tools for promotion of science and technology development have to be directed to the development of creative potential,
- Basic criteria for development of creative potential is its comparison with the world standard.

Within the framework of these principles we consider technological education as a necessary presumption for the fulfilment of not only our development strategy for science and technology, but also for the development of economy as a whole. The necessary presumption for the realization of our scientific and technological policy in the benefit of development of economy, is most of all the education of employees, present and future managers, and pro-innovation thinking and acting. In this direction we are facing a task to overcome consequences of the past directive economic regime, which did not allow preparation of creative and managing people in a business oriented approach to an innovative behaviour.

And right in this specific need of our economy, in the need for quick supplement of theoretical and practical knowledge in innovation technologies as resources for development of business activities, we see the meaning and possible benefit of your conference.

I wish you success in your effort to achieve results which will facilitate the fulfilment of our needs.

The Development of Technology Education in New Zealand

Don Ferguson
New Zealand Ministry of Education
Department of Policy
P.O. Box 1666, Wellington, New Zealand

Abstract. The paper firstly reviews changes to the structure of education in New Zealand, and discusses their influence on the decision to include technology in the curriculum as an essential learning area. The challenge of introducing a new area into the curriculum is described in terms of six critical issues. The approach to the development of the draft curriculum statement for technology, including consultation strategies and the development of a framework for technology education, is outlined. The draft curriculum statement is described in terms of three general aims and six strands, six contexts, and seven technological areas. Government policy for teacher development in technology, funding for facilities and equipment, and the importance of schools establishing links with the community is discussed. The paper concludes by describing how schools are proposing to deliver the technology curriculum.

Keywords. Curriculum reforms, structure of education, technology curriculum, technology education

1 Reforms in New Zealand Education

In 1984 major economic reforms began in New Zealand. In the decade since then New Zealand has moved from a very regulated economy to a free market economy. Over the last seven years, education in New Zealand has undergone fundamental and wide-ranging change. The purposes of the reforms were to make education more responsive to the needs of the community and to make schools more accountable for the way they used publicly owned resources.

The key common principles in the reform of all sectors include:

i. Charters
ii. Devolution
iii. Bulk Funding
iv. Review.

2 Educational Reforms and Curriculum Development

As part of the reforms in education, the Curriculum Development Division of the former Department of Education was dissolved. Curriculum development is now undertaken in a different way - by contractual arrangements. The Ministry of Education has a Learning and Evaluation Policy section responsible for developing policies for the development of the New Zealand Curriculum, and a Curriculum and Contracts Management Section responsible for the implementation of curriculum policy.

3 Curriculum Change

During the 1980's there were calls for a more coherent curriculum policy. Following its inception in 1989, the Ministry of Education began developing a comprehensive framework for a new New Zealand Curriculum.

In May 1991 a draft document The National Curriculum of New Zealand was issued to schools for comment. This statement proposed a comprehensive national curriculum framework for the development of policies for teaching and learning in New Zealand schools. After widespread consultation this draft document was revised and published in May 1993 as the New Zealand Curriculum Framework.

The New Zealand Curriculum Framework is the policy statement for teaching, learning and assessment. This statement and the associated national curriculum statements[1] set national directions in education and provide a basis for clearly indicating to parents, students, teachers and the wider community what should be taught and learned during the compulsory period of schooling.

The framework has the following elements:

– Principles - the broad curriculum fundamentals
– Essential Learning Areas - the broad aspects of knowledge and understandings for all students
– Essential Skills - the skills and qualities for all students
– Attitudes and Values
– Assessment

---

[1] There is a national curriculum statement for each of the seven essential learning areas.

One of the seven essential areas of learning described in the New Zealand Curriculum Framework, and a new area of learning for New Zealand schools, is technology.

4 Background to the Development of the Curriculum Statement for Technolgy

4.1 Research in Technology Education

In 1989 a research project was set up to study students' concepts of technology and attitudes to technology. The research instruments used were those developed at the Eindhoven University of Technology (as part of the PATT project) in The Netherlands and used in many countries. The PATT tests, together with an open-ended question, were administered to a national sample of year 8 students.

It was found that New Zealand students had poor concepts of technology but positive attitudes to technology. Many students perceived technology to be the physical products of recent developments such as computers and high technology equipment, and of overall benefit to human beings. Concepts were positively related to attitudes.

It was decided in 1991 that further developments in technology education needed to be based on research about how students develop technological skills and understandings. A survey of overseas literature revealed that there had been little research in this area. As a result a contract for a three year research project, Learning in Technology Education (LITE), funded by the Ministry of Education, was let to the Science and Mathematics Research Centre at the University of Waikato. This research is still continuing and has been concerned principally with developing student's technological capability. Reports published to date are:

- Teachers' perceptions of technology education
- Development and management of technological capability
- Analysis of student technological capability
- Working with teachers to enhance student technological capability.

Information from this project is providing valuable advice to the Ministry of Education about the development of technology education, and to teachers about learning in technology and appropriate strategies for the teaching of technology.

4.2 Policy Development

In 1991 a Ministerial Task Group was set up to review science and technology education. Their report entitled Charting the Course, released in January 1992, supported technology education being included in the curriculum for all students. The Task Group made suggestions about the nature, scope, and delivery of technology education within the New Zealand context.

During 1991 a major literature search was undertaken by officers in the Ministry of Education. Discussion papers arising from the literature search and review of overseas curriculum documents were made available and have proved to be helpful to teachers and educators seeking further information.

In November 1991 the Learning Media section of the Ministry of Education distributed a video to schools (Education Update Number 8) showing some current technology initiatives in schools; and in March 1992 a discussion booklet "So This is Technology" was issued to schools based on teachers' suggestions and questions. Both items were designed to encourage discussion about technology. education.

During 1992 seven policy papers were prepared for the Minister of Education outlining options with regard to the nature, scope and delivery of technology education.

Six factors were identified as important to the success of the development and implementation of the technology curriculum. They were:

– A framework which schools could implement in a flexible way and which took account of the existing ideas of teachers and the realities of schools;
– Involving as many teachers as possible in developing and commenting on the framework;
– Support from school boards of trustees and the community;
– Adequate teacher professional development;
– Adequate resourcing for teaching materials and equipment;
– Links with appropriate tertiary education groups and business and industry.

These issues are addressed later in the paper.

The policy papers formed the basis of a definitive discussion document, "Technology in Schools" distributed to schools in May 1993. Following consideration of comments, and widespread consultation with the education, business and industry communities, a contract was let to the Mathematics and Science Education Research Centre at the University of Waikato to develop a draft curriculum statement for technology education.

5 The Framework for Technolgy

The contract for the development of the curriculum framework for technology was managed by Dr. Alister Jones. Over 200 teachers and were involved in the preparation of the statement, as well as people from business and industry. This document was distributed to schools for trial and comment in December 1993, and all teachers have been invited to comment on the draft statement.

The curriculum statement gives a brief rationale for technology being introduced as an essential learning area and suggests three aims for technology education:

– Development of technological knowledge and understanding;
– Development of technological capability;
– Development of awareness and understanding of the relationship of technology and society.

These three aims lead directly to the six inter-related learning strands and their associated achievement objectives. The strands are:

– Technological knowledge and understanding;
– Identification of needs and opportunities;
– Implementation and production of technological solutions;
– Communication and presentation of strategies and outcomes;
– Reflection and evaluation;
– Technology and society.

The curriculum statement specifies seven technological areas. The technological areas are as follows:

- **biotechnology** includes the utilization of living systems for the development of processes and products to benefit people;
- **information and communications technology** includes systems that enable the acquisition, manipulation retrieval and communication of information in various forms;
- **electronic technology and control technology** includes the design, construction, and production of systems and devices, from simple electrical circuits through to integrated circuits, robotics, and control system;
- **food technology** includes the safe production and processing of food and the development, packaging, and marketing of foods;
- **process technology** includes chemical, industrial, manufacturing, and transportation processe;
- **materials technology** includes the use and development of materials to achieve a desired result; knowledge of different types of materials; and the processing, preservation, and recycling of materials;
- **design and graphics technology** includes the uses of different materials, graphics, and modelling to develop designs and to communicate ideas and technical information.

Possible learning experiences and assessment examples are listed for each strand. Most technological activities which students undertake will address all, or a number, of these strands.

Timetable for further developments:

- Responses to the draft curriculum are due by 31 December 1994
- Analysis of the responses and revision of the curriculum statement - Jan/Feb 1995
- Distribution of the final curriculum statement to schools - August 1995
- Official implementation of the statement from the beginning of 1997

6 Support from Teachers, School Boards of Trustees, and the Community

This issue has been addressed in two ways. Firstly through publications designed to inform teachers, school boards of trustees, and the community about the rationale for inclusion of technology in the curriculum and about the nature and scope of the technology curriculum. Two publications and a video tape have been distributed. Schools also receive a regular Technology Update which is a progress report on Ministry of Education technology developments.

Secondly a television programme 'Know How' has been produced and has recently been screened by Television New Zealand. Each episode of the ten part programme was screened on a Tuesday morning and repeated on the following Saturday. Titles of the programmes are:

- The Call for Technology Education
- The Aims of Technology Education
- The Technology Curriculum
- The Curriculum in Action I
- The Curriculum in Action II
- The Curriculum in Action III
- The Question of Assessment
- Implementing Technology Education
- Resources and Facilities
- Community Links and Overview.

While many schools viewed the programmes as they were broadcast, and others taped them, copies of the programmes are being made available to all schools. It is proposed to repackage this series for various purposes including as a distance education technology education course.

7 Teacher Professional Development

The Learning in Technology project has provided input into the nature and scope of the professional development needed for technology. In the recent Budget the Government announced a budget of $11 million over the next three years for teacher development for technology. It is likely the following model will be adopted:

Phase 1   Development of models for teacher training
          Training of trainers

Phase 2   National Teacher Development Programmes
          Whole-school based
          Open Learning Packages
          Advanced Studies for Teachers (AST) papers

Universities are already offering graduate and undergraduate papers in technology education both in the philosophical base for technology education and in the specific technological areas.

8 Equipment and Facilities

Schools now have much of the equipment to implement the new technology curriculum. However a policy paper was commissioned on facilities and equipment for technology and the need for some additional resources was identified in the paper. Schools in New Zealand are bulk funded through the Operational Grant to Schools. In the recent budget an additional $6.4 million per year was added to the Operational Grant for implementation of the technology curriculum. Also as technical and home economics facilities become eligible for upgrading they will be upgraded as technology centres. Upgrading of these facilities has a priority in the National Capital Works Programme for schools.

9 Links with Industry

This is considered to be particularly important and is being addressed in several contexts. The Minister's Policy Advisory Group for Technology includes industry representatives. People from business and industry and university engineering and technology departments have been involved in the development of the draft technology curriculum statement. The Ministry of Research, Science and Technology are making available scholarships for teachers to have time out in industry, and the New Zealand Education Business Trust is working to match all secondary schools with a local business/industry.

Other industry groups are sponsoring technological activities for students beyond the classroom such as Crest and the BP Technology Challenge. These events are involving tens of thousands of students working as individuals or groups in technology problem solving activities.

10 Implementation of the Technology Curriculum

It is expected that schools will implement the technology curriculum in a variety of ways. Electronic, print-based and face-to-face mechanisms for sharing implementation strategies through case studies and other approaches are planned.

Approach at primary schools. Schools have the flexibility to decide how they will implement the technology curriculum. In primary schools it is expected that technology will be implemented through an integrated studies approach, e.g., making a device to tell when a bird lands on a bird table as part of a science unit on birds, making a database of employment in the local community as part of a social studies unit. However in the later years of primary school there are also likely to be some discrete studies e.g., designing and making an automatic bird feeder, writing a programme to control a robotic crane.

Approach at secondary schools. Again schools have the flexibility to decide how they will implement the technology curriculum. Many schools have already indicated that they will adopt a cross-curricula approach. Some schools have indicated that they will develop a subject called technology and a few schools will do both of these. What they decide to do will be influenced by their understanding of the concept of technology education, the background and skills of teaching staff, resourcing issues and timetabling. It has been strongly recommended to schools that a technology coordinator be appointed from senior management so that it is not captured by any one subject area.

In the senior secondary school it is likely that many schools will offer separate courses related to the various technological areas. A bursary technology examination is planned with students being able to elect an area/s of technology they will be examined in.

11 Conclusion

The past five years have been a time of great change in New Zealand education. A new administrative structure has been put in place and major curriculum reforms, involving the total curriculum, are under way.

The importance of including technology as an essential area in the curriculum has been recognized through its inclusion in The New Zealand Curriculum Framework and through the provision of funding for teacher professional development in technology and other resources. The challenges ahead include developing teacher professional development programmes that will give teachers the knowledge and confidence to undertake technology in their classrooms, convincing teachers of the benefits of being prepared to work with colleagues in designing and presentation of teaching programmes, and developing acceptable assessment strategies.

References

1. Jones, A., Carr, M.: Teachers Perceptions of Technology. Centre for Science and Mathematics Education Research. University of Waikato, Hamilton 1992
2. Jones, A., Carr, M.: Development and Management of Technological Activities. Centre for Science and Mathematics Education Research. University of Waikato, Hamilton 1993
3. Jones, A., Carr, M.: Analysis of Student Technological Capability. Centre for Science and Mathematics Education Research. University of Waikato, Hamilton 1993
4. Jones, A., Carr, M.: Working with Teachers to Enhance Student Technological Capability. Centre for Science and Mathematics Education Research. University of Waikato, Hamilton 1993
5. Ministry of Education: The New Zealand Curriculum Framework. Wellington 1993
6. Ministry of Education: Technology in Schools. Wellington 1993
7. Ministry of Education: Technology in the New Zealand Curriculum. Wellington 1993
8. Ministry of Education: Policy Papers on Technology in the Curriculum. Wellington 1993

Technology Education in Germany's Gymnasia

W. E. Theuerkauf, W. Haupt, W. Wagner, A. Weiner
Institute of Applied Electrical Engineering and Technology Education
University of Hildesheim
Kreuzstraße 8, D-31134 Hildesheim, Germany

Abstract. The school system of the Federal Republic of Germany differs from that in the Anglo-American world in the way it is structured, offering as one characteristic feature the Gymnasium. To be able to understand the Gymnasium with its specific tasks as part of the entire school system, the organization of this system and the different types of schools with the tasks and objectives assigned to them have to be illustrated so that from the overall picture the specific task of the Gymnasium and its educational mandate can be derived.

The general educational mandate will be illustrated as well as how general technology education is dealt with in the final years Gymnasium (or senior high-school), reference being made to syllabus and the methodology applied. The dividing line but also the links existing in this context between Gymnasium and the wide-ranging and comprehensive technology education as part of vocational training will be shown. Another aspect is how at this school level the connection is with regard to contents and methodology established between technology as a subject with the metascience of 'general technology' and technology as a science. This science-related theoretical basis justifies the technical orientation of the course system in the senior grades of a Gymnasium. In conclusion, an example is to illustrate how the objectives of technology education outlined are translated into actual teaching at the Gymnasium.

Keywords. Technology education, general technology, gymnasium, curriculum

1 The Structure of the Educational System in the Federal Republic of Germany

The educational system in Germany encompasses the two sectors of a general all-round education and of vocational training. Having been formed in the course of several hundred years and influenced by Humboldt's theory of education, the former aims to provide the cultural techniques, but also to equip pupils with an encompassing and broadly-based knowledge. It finds its concrete shape in dealing with and developing patterns of action that will help develop ones personality and master their lives at present and in the future. Vocational training on the other hand, is to impart the necessary abilities within a defined field, which can either be a trade or a specific profession. It follows from the above that the two sectors within the German educational system have to differ in their curricular and educational approach. In this connection it should be underlined that, although the two sectors are of equal standing, they are not identical. They are interdependent and they supplement each other in their aspiration to help a young person develop his or her own and comprehensive personality.

Changed job requirements as a result of new ways of work organization and the growing role of key qualifications (Theuerkauf, Weiner 1991) have blurred the dividing line between the methodologies (development of competencies of action and decision) applied in the two sectors. After all also vocational training is obliged to convey not only specifically technical but also multi-functional thinking and acting.

For the Gymnasium, which, though in a modified way, still follows the historically developed ideal of the humanities, the distinction into general education and vocational training has - in the way it sees itself - not lost significance. Quoting Humboldt, the classical ideal is an education keeping a critical distance from trades and professions, economy, state and society. The neo-classical Gymnasium, which to the present day for the general public has remained the 'classical' Gymnasium, had been formed against the immediate background of the 1810/1812 reforms in Prussia and had been given three principle pillars. Firstly, the examination pro facultate docendi which at the same time created a specific teaching profession for the Gymnasium independent of the ecclesiastical office and of any theological studies. Secondly, the curriculum concentrating on the four main subjects Greek, Latin, German and Mathematics, and thirdly the final examination of the Abitur allowing those leaving the Gymnasium, and only those, to take up any course of studies at university. In the year 1900, three types of nine-grade secondary schools were given equal status: the classical Gymnasium teaching both ancient languages, the Realgymnasium that only retained Latin, and the Oberrealschule that did not include Latin. Today, all the secondary schools carry the name Gymnasium, the version teaching ancient languages being regarded as the exception. The classical principle, however, still prevails. For the natural sciences and modern languages that the needs of a society in our industrial era have placed into the hands of school and the training provided there could derive their pedagogical justification only from the old classical approach (Blankertz 1969). When viewed against this basic conception, the educational value of technology appears to be questionable, and technical subjects would consequently not find a place within the canon of subjects taught at a Gymnasium.

This conception produces a school system where general education and vocational training are, with a few exceptions, taken care of by different educational institutions. These two part-systems do, however, not exist as isolated units within the overall system. They allow for enough openings through which pupils can change from one system to the other, the only provision being that they have the required entrance qualifications.

Without going into the particulars of the system with its many diversifications, the organizational structures of the educational system shall be expounded below. Having started with kindergarten, any child has to attend school at the primary or elementary level. The next level is the main school (Hauptschule), the middle school (Realschule) or the Gymnasium[1]. The school leaving certificate for pupils finishing school after the main and middle schools is that of Sekundarstufe I, while the Gymnasium is completed with that of Sekundarstufe II or the Abitur. After the secondary grades follows the professional or vocational training.

School-leaving examinations are always understood to be an entrance qualification for the next higher level in the educational system. To have passed the Sekundarstufe I examination enables a pupil to continue his or her education at Sekundarstufe II, and the final examination at this level, the Abitur, qualifies this pupil to take up any subject at university. In practice, this general university entrance qualification is restricted by the so-called numerus clausus. The final examination standards are defined by a regulatory framework that is agreed for and by the different German Länder and set forth in examination regulations and directives. To make sure that standards are met, some of the German Länder have a central final Sekundarstufe II examination.

Sekundarstufe I and II lead up to professional or vocational training, for which schools providing a general education prepare. Pupils leaving the main school or the middle school will normally take up a vocational training, with main-school leavers tending to favour practice-oriented jobs, and those having completed middle school normally preferring jobs making higher demands on their theoretical abilities. Vocational training is as a general rule provided through a dual system with the state-run vocational schools on the one hand and operations in industry, commerce and trade on the other (Theuerkauf, Weiner 1993). The primary objective of main schools and middle schools thus inevitably is to prepare pupils for vocational training and ultimately their jobs in an actual working environment.

At a Gymnasium, the majority of pupils pass through the Sekundarstufe I and II levels. At the Sekundarstufe II level, they have achieved their 'Abitur', which at the same time is the required entrance qualification for university. The Sekundarstufe II level, which is equivalent to the final years at a Gymnasium, thus prepares for university. At this level, it is attempted to impart a certain propaedeutic approach in subjects such as German, mathematics, the natural sciences and foreign languages, from which university curricula will start.

---

[1] In Germany, there are in addition the comprehensive schools combining within themselves the main school, middle school and Gymnasium levels.

Technical Education and the Connection to the System of Education

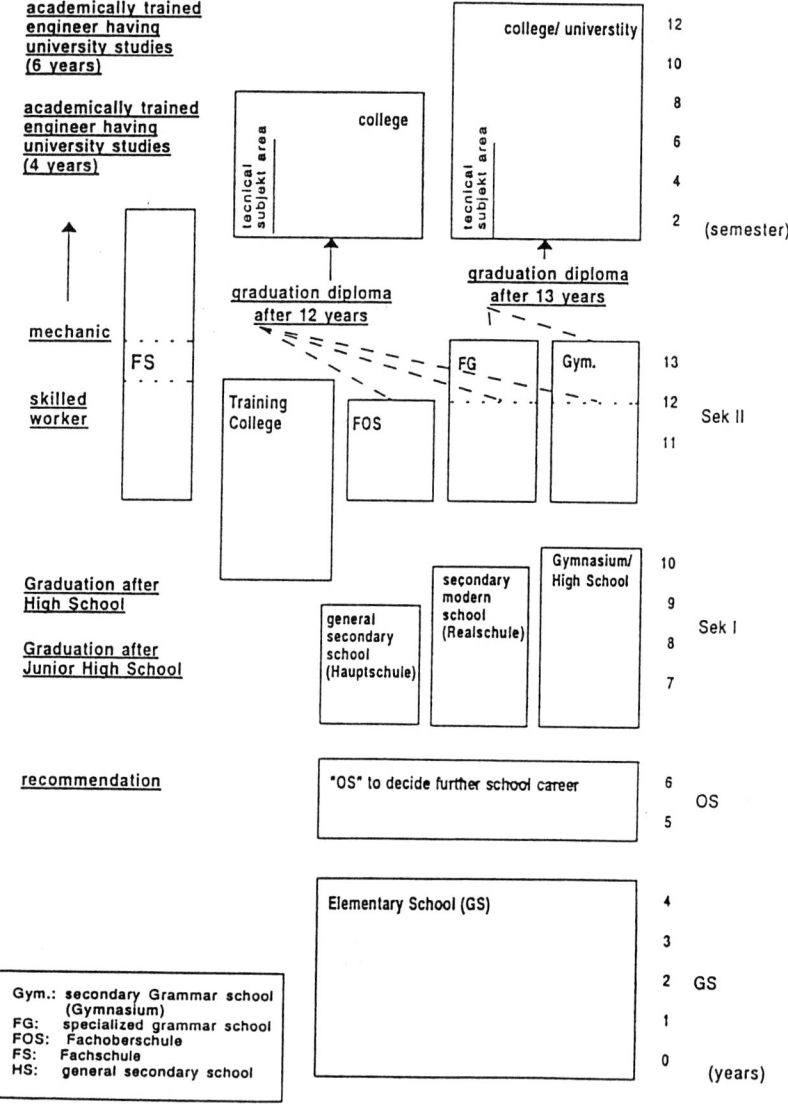

Fig. 1 Structure of the educational system in Germany

Technology education is present at all levels of school education and all types of schools providing a general education. At elementary schools, basic education in technical matters forms part of general studies (Sachunterricht) where in a phenomenological approach direct reference is made to the environment as experienced by the child. During the first two years at the Sekundarstufe I level, technology teaching centres around handicraft skills with an emphasis on the ability to handle materials and simple tools. In the following years, there is a shift to a basic technology education, turning for the first time also to specific professions or trades[2] (Theuerkauf 1983). In conjunction with the subject 'economics' (Wirtschaft), the emphasis of basic technology education is on the technical structures determining both professional life and leisure. Vocational training makes a four-week training course in industry obligatory. Compulsory education ends with the Sekundarstufe I level.

Sekundarstufe II offers, on the one hand, a broadly based technical vocational training that can be terminated as a skilled worker or continued to become a master craftsman (Meister) or a technician. On the other hand, there is the Gymnasium offering general technology education. The latter aspect is to be dealt with in more detail below.

2 Objectives for the Final Years at a Gymnasium

The parallelism that exists between the technical vocational training and the technology education as part of general education at the Sekundarstufe II level directs the attention to the question of the right of the latter to exist and thus in how far technology education within the canon of Gymnasium subjects is a general-knowledge subject. Of all the Federal German Länder, only North Rhine-Westphalia has settled this basic question in favour of technology teaching.

As a new subject, technology education has to start from the objectives governing the traditional Gymnasium subjects and forming the underlying principle of school work during the final years of a Gymnasium. These objectives are defined twofold: by propaedeutic instruction and methods and by the assistance pupils are to be given for self-fulfilment and social responsibility (Minister of Education of North Rhine-Westphalia, 1981).

Propaedeutic instruction requires:

– that the principles and possibilities of independent work are mastered in general. The solution of problems and the ability to reflect circumstances and form a judgement on the basis of sound knowledge should be mentioned here as one characteristic element.

---

[2] The terminology for the subject imparting technology at varying degrees of emphasis may vary from on German Land to the next. At the primary level (up to grade 7) it may be termed 'Werken' (arts and crafts), while for Sekundarstufe I (grade 8 plus) this may be 'Arbeitslehre' (industrial science) or 'Arbeit/Wirtschaft/Technik' (crafts - economics - technology).

– the ability to practice the fundamental scientific procedures and methods of finding, in order to be able to recognize and identify structures and methods in science and evaluate their validity in particular in complex systems.

This objective for education before finishing the Gymnasium, which ranks above the different disciplines and sciences, has been transformed to the level of specific scientific practice and methods of finding. At this school level, translation into teaching practice can only proceed by way of example, and this refers to any of the subjects taught.

The objective of self-fulfilment as a major element in education, which is characterized by the ability and readiness to communicate and work in teams, corresponds to the propaedeutic emphasis. Where due consideration is to be given to social responsibility, both objectives can be achieved only when at the same time concentrating on humanitarian values or a system of values based on society (Rekus 1991).

3 Aims and Objectives of Technology Education

For pupils in their final years at Gymnasium, there is a choice of a number of major areas, which follow the different academic specializations:

- classical philology
- modern languages
- mathematics and natural science
- economics
- social sciences
- technology
- music
- sports.

The mathematical/scientific/technical branch includes as subjects mathematics, physics, biology, chemistry, informatics and technology. Some of the subjects in particular foster insights into technical contexts. Technical systems for instance can be described with the aid of mathematical relationships, while for an analysis and synthesis of technical artefacts and processes one very often has to rely on physical laws. These requirements correlate with the conditions engineering students encounter at university, where again they have to prove that they possess the basic mathematical and physical knowledge and abilities.

For curricular decisions, the technology definition according to Huning (1974) can be used, who defines technical practice as a purposeful changing of nature by material means under specific economic and social conditions. Wiener (1963) refers to the categories matter, energy and information to describe technology. Within the artefacts and processes of the different technical disciplines, these undergo purposeful changes that are brought about by creative design and its realization through the production process, due consideration being given to the prevailing economic and social conditions.

General Technology (Beckmann 1806; Wolffgram 1978) as the theory of technics integrates the different disciplines of engineering. Technology as a science can thus be defined as a theory of cognition and methodology similar to the traditional sciences. It thus represents an instrument of cognition as it produces the links and relationships between the different technical disciplines, e.g. on the basis of a comprehensive understanding of systems and processes (figure 3). It can be expanded into a model of hierarchies, which horizontally covers interdisciplinary structures and vertically gives concrete shape to concrete technical realization (Dyrenfurth 1990).

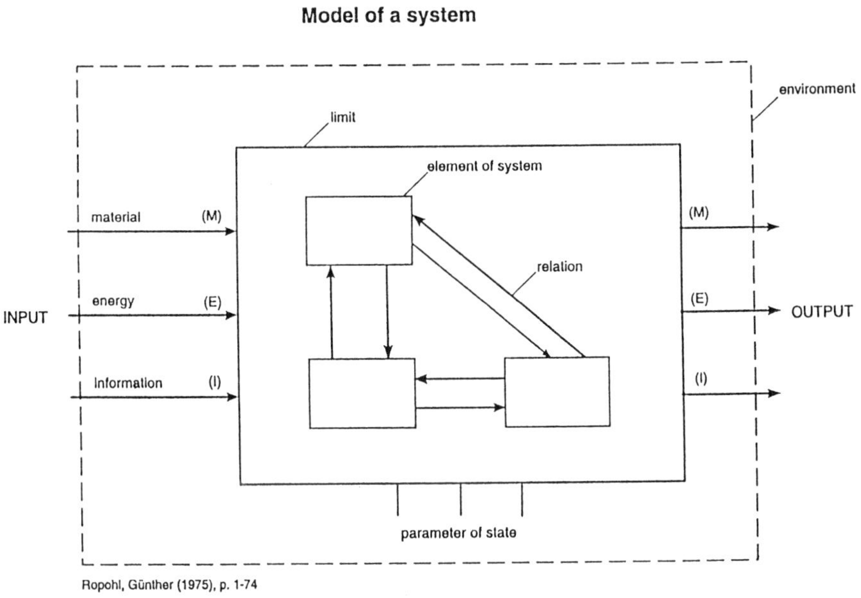

Model of a system

Ropohl, Günther (1975), p. 1-74

Fig. 2 System model

For a description of technical objects and processes in particular for technology teaching the systems theory is a useful instrument (Theuerkauf 1983). It allows the structures of elements, or the patterns of relationships, to be described at different levels of concretization. The elements, which are characterized by their function, are determined by the relationship between input and output on the one hand, and by the functional response on the other. The functions are time-dependent and can be expressed as mathematical functions. Logic linking of the elements, which can be a serial, parallel or retroactive linking, produces a network topology. From this network topology, the structure of the technical object can be derived as a system. The process represents the time dependence of the overall system; it is progress-oriented and can be described by means of an algorithm.

The purposeful changing of material means in a narrower sense can be understood to be the design in as much as it is a technical process of creating and shaping, which always regards itself as a process of optimization, that is of development, progress and improvement. This process integrates:

- system organization: the definition of a duties record book, due consideration being given to socio-technical aspects
- system development: the geometric forming of artefacts and processes
- system production: the preparation of engineering data and the logistic organization of the production process, including quality assurance
- system utilization/system recycling: system analysis, also with a view to economic and ecological aspects.

A complex approach for the purposes of General Technology is represented by a spatial system of coordinates entered into which are the different categories (figure 3). One obtains on the one hand the operands matter, energy, information with their operations, and on the other the socio-technical interface.

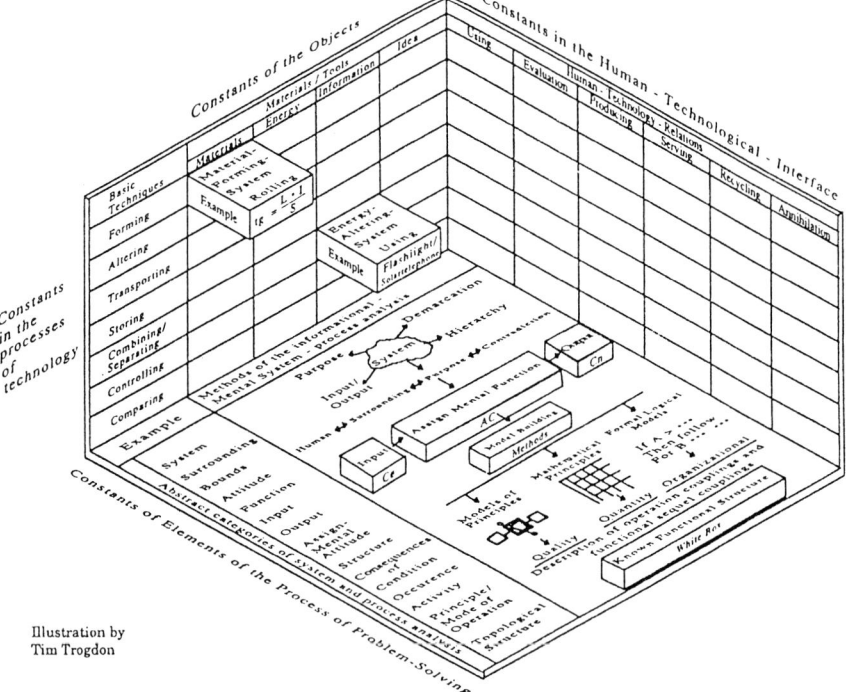

Illustration by
Tim Trogdon

Fig. 3 Model planes. Constants of technological processes vs objects vs interaction sites (Blandow 1991)

By linking these variants, a point is defined that describes the degree of examplarity teaching items have to possess. Thus, a morphological field is marked out that can also be applied to the scope of curricula in technology education.

From this scientific and theoretical approach, which permits technology as a science and its structures to be defined, the objectives of technology teaching in Gymnasia were outlined (Minister of Education, North Rhine-Westphalia 1981):

- acquisition of technological fundamentals and technical knowledge
- introduction into technical thinking and acting
- fundamentals of adequate, reflected and responsible reaction in situations where technical aspects are one determining factor.

Although a clear-cut differentiation will be dispensed with here, the objectives are interpreted as follows:

The acquisition of technological fundamentals and technical knowledge is in particular to form the basis for the development of the ability to furnish an analysis/ interpretation and a synthesis/description of specific and complex systems and processes, including but not being limited to those in power engineering, production engineering or data systems technology. This includes the proper use of the relevant terminology and adequate transformation of these model descriptions.

Scientific thinking and acting in a technical context means on the one hand to be able to draw on specifically technical experimental methods to understand and judge functional relationships, and on the other to initiate solutions to problems (on a school level) in line with the design methodology, that is by developing, progressing, improving. This implies at the same time that the dimensions of the individual and of society are integrated.

Adequate, reflected and responsible reaction in situations affected by technical circumstances means to be able to give a scientifically based appraisal of the validity of technical conditions and the actual significance of technical systems and processes for society. To be able to arrive at an adequate decision, the historical development and, on the basis of the socio-technical models, the methods of Technology Assessment however need to be taken into account.

4 Technology Curricula Within the Course System of the German Gymnasia

Technology as a subject was introduced in the final years of the German Gymnasium in the year 1981[3,4]. Its basic principles go back to Huning's

---

[3] The curriculum concept was developed by working groups at the German universities of Essen and Duisburg

[4] Special credit for the concept and the political implementation and introduction of the curriculum is due to Prof. Dr.-Ing. W. Wagener.

definition of technology. The course system starts from the comprehensive model in figure 2, in which systems of a matter, energy and information transfer interact as a combination of technical systems, an interaction also taking place with the social, economic and ecological surroundings. Technology as a required elective subject has to follow the requirements that are also applicable for other sectors at this school level. It is taught for six 6-month courses, each course covering three class periods per week.

Technology courses at the advanced level of the gymnasium

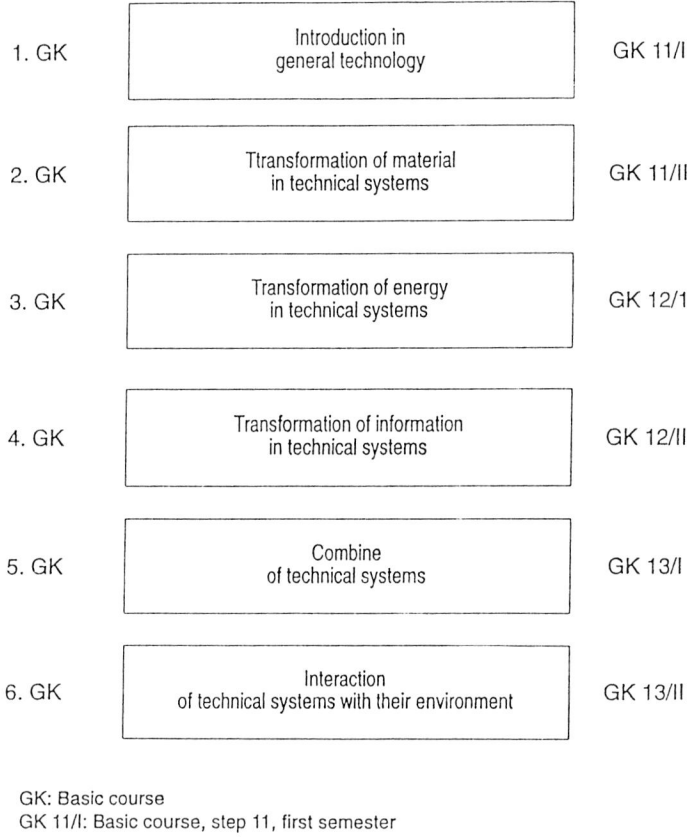

GK: Basic course
GK 11/I: Basic course, step 11, first semester

Kultusminister NRW (Hrsg.): Richtlinien für die gymnasiale Oberstufe in Nordrhein-Westfalen. Köln 1981. S. 39.

Fig. 4 Technology courses in the final years of a Gymnasium

Directives give more concrete shape to this general course structure (Minister of Education North Rhine-Westphalia 1981, p. 39) with the aid of the example of the system of energy supply running like a thread through all the six courses. Following the principle of a comprehensive technological approach, each course concentrates on a different aspect, or a different part-system, of this global system. Course 11.2[5], for instance, deals with the refinery as a system in which a conversion of matter takes place, reference being made to the processing of crude oil into fuel oil. For course 12.1, the conventional power station heated by fuel oil serves as an example of the energy aspect as an overall system. Course 12.2 concentrates on the flow of information within a power station, that is on controlling and automation, course 13.1 on the network interlinking energy stations. Course 13.2 in conclusion tries to highlight the interrelationships between the system of electric power supply and other systems.

This general framework also permits of other general issues as course subjects. These could be networked production systems, environmentally acceptable traffic or communication systems, or similar subjects comprising complex systems. To comply with the comprehensive approach, any such subject however would, have to be dealt with as a combination of technical systems integrating as part-systems the aspects of matter/energy/information conversion and the socio-technical aspect.

Technical competency to act, which is also referred to as technical thinking and acting, is imparted through a combination of subject-specific and comprehensive teaching methods. These take account of the pupils at which they are directed and the propaedeutic approach, but also the school situation.

Starting from these directives, teaching methods can be structured as follows (Minister of Education of North Rhine-Westphalia 1981, p. 75):

- subject-specific methods, including for instance technical experimentation, synthesis of technical system components by way of calculation and design, optimization, analysis of technical systems
- comprehensive teaching methods, including for instance project tasks, simulation games, case studies, scenario methods.

5 Teaching Practice

The framework outlined above leaves the instructor enough room for manoeuvre to translate the relatively abstract educational objectives into an actual syllabus. The example "energy production" is to illustrate how the concept is translated into practical teaching.

For each course, the guidelines provide sub-titles which in turn are subdivided into teaching items. The sub-titles and teaching items result from or are determined by the actual course subject. The course subject of course 12.1 for instance is the energetic aspect of the power station, which will be elucidated in

---
[5] 11.2 means year 11, course 2

particular by the system components and the power station circuit diagram (figure 5), as well as by the different stages of energy conversion within the power station.

Components of system of a conventional power station and steps of power conversion

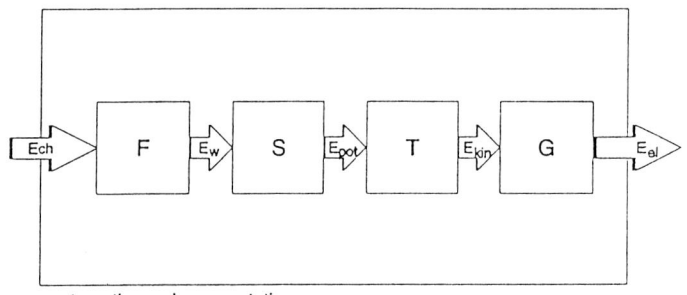

system: thermal power station
F: furnace
S: steam generator
T: turbine
G: generator

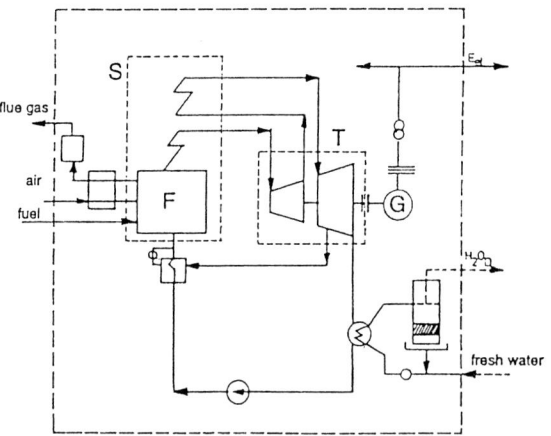

simplified wiring scheme of a steam power plant

Wagener, Willi (1985), p. 9, 35.

Fig. 5 System components of a conventional power station and stages of energy conversion

A system-related approach produces the following sub-titles:

1. Objectives of systems of energy conversion - techno-historical development
2. Principles of energy conversion
3. Realization of energy conversion in technical systems
4. Energy transmission within technical systems of energy conversion
5. Elementary diagram of a complex system of energy conversion
6. Optimization of the thermodynamic cycle
7. Analysis of a system of energy conversion
8. Linking a complex system of energy conversion with its environment, and assessment.

Subtitles (3) and (8) - 'Realization of energy conversion in technical systems' and 'Linking a complex system of energy conversion with its environment and assessment' - will now be used to illustrate how the course subject 'turbine' can be covered in class. The turbine is in this connection considered as one stage within the energy conversion system and thus as a power station sub-system (see figure 5). The teaching aids for this course (Wagener, Schmidt 1981) show how the turbine can be used for technical experimentation.

With such an experiment, parameters and ways of optimization are analysed, such analysing in turn producing for instance hypotheses; a number of measuring methods can be selected, graphs can be produced. The findings can, moreover, be integrated into complex system patterns, and conclusions can be drawn as to the design of specific items. To allow this experiment with the objectives outlined above to be carried out at schools, a classroom model has been developed (see figure 6).

For reasons of safety, this model uses compressed air instead of steam. The model will illustrate the nozzle jet angles and the rotor blade angles, the optimization experiment consisting in a variation of these parameters so that the turbine achieves the highest possible degree of power transformation (see figure 7).

Subtitle (8) - Linking a complex system of energy conversion with its environment and assessment - can through the example of the turbine be given concrete form. This can for instance be done by showing how the improved efficiency of a power station can reduce the environmental hazards produced by CO_2 emissions. Since one factor affecting the efficiency of a power station is the efficiency of the turbine, this experiment can serve as a good example for the close interrelationship between improvements in the technical standards and ecological systems. An analysis of the second law of thermodynamics furthermore reveals, again through the example of the turbine, that to raise the steam temperature could also be a way of improving the efficiency. This measure, however, presupposes that materials are developed for the turbine that meet the requirements at elevated temperatures without impairing their mechanical properties.

Experimental model of a turbine

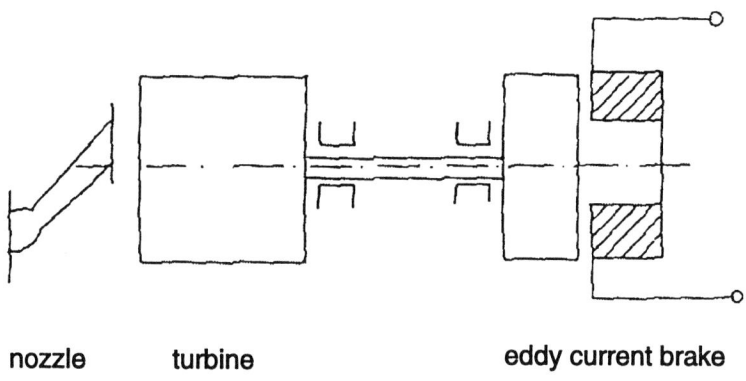

nozzle   turbine   eddy current brake

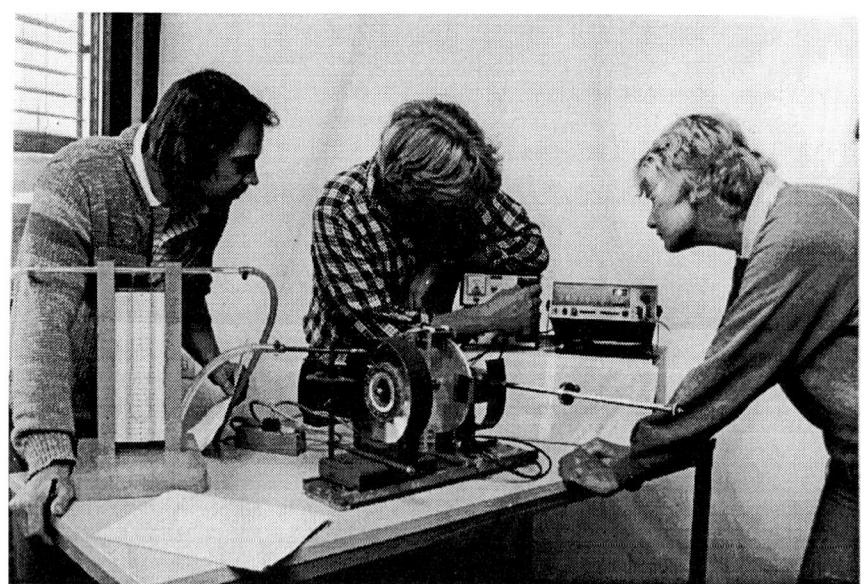

(Wagener, Schmidt 1981)

Fig. 6 Classroom model of a turbine

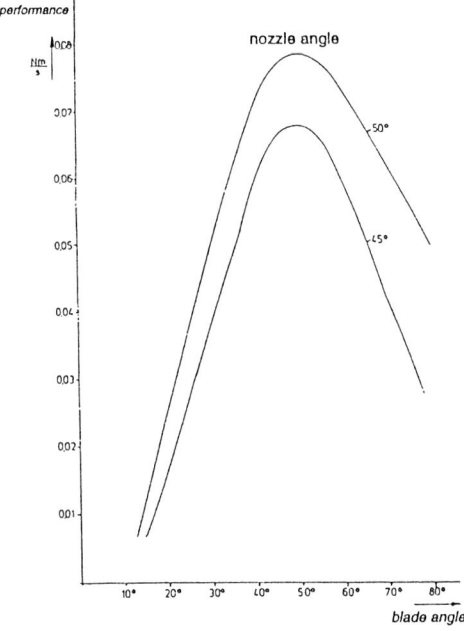

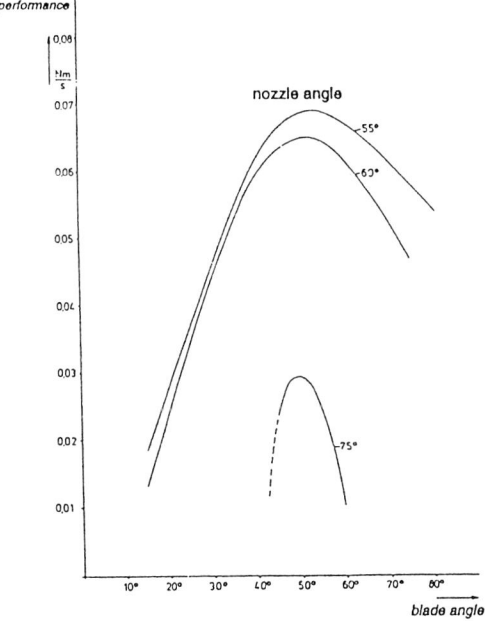

Fig. 7 Turbine performance as a function of the blade angles and different nozzle settings

6 Summary

Technology education at the Gymnasium is more than just propaedeutics for university studies in natural sciences or engineering. This can be explained by the fact that it is not immediately related to a specific discipline of engineering, but tries to provide all the pupils with a guideline and a basis for judgement in a world in which technical matters are one major factor. Technology education as part of a general-knowledge education thus clearly differs from the specialized technology education as part of vocational training. General technology as a science of technics thus complies with the educational objectives in the final years of a Gymansium, which are directed at scientific propaedeutics. Having been recognized as an engineering science, it permits of a differentiated description and hence disclosure of highly complex technical structures.

Technology education at schools providing a general education cannot restrict itself to a technical orientation, it always has also to include the human factor. Technical action as one of the factors giving shape to the social environment of the individual hence has to provide a link with ethics when part of school education. The systems theory allows technology and environment to be brought into a relationship, and in conjunction with empirical methods (Steffen, Weiler 1991) it provides the basis on which such socio-technical issues as the employment situation and the sources of energy can be treated by scientific methods of Technology Assessment and thus to arrive at an appropriate judgement.

Since its introduction as a subject at about 30% of the Gymnasia in North Rhine-Westphalia, despite strong resistance technology teaching has held its own on the basis of this concept and it has proved to be an attractive subject for pupils that deserves to be treated on an equal footing with other subjects. Committed teachers, including university teachers, and encouragement from industry provide the stimulus for development of this curricular approach.

References

1. Beckmann, J.: Entwurf einer allgemeinen Technologie. Göttingen 1806
2. Blandow, D.: The Elements of Technology Education. TU Eindhoven 1993
3. Blankertz, H.: Bildung im Zeitalter der großen Industrie. Hanover 1969
4. Dyrenfurth, M. J.: Rethinking Technology Education in the Secondary School: Missouri's Approach to Technological Literacy. Erfurt 1990
5. Haupt, W., Sanfleber, M. representing the Essen and Duisburg working groups: Ansatz zu einer Didaktik der Technik für den Technikunterricht in der Sekundarstufe II (differenzierte gymnasiale Oberstufe). In: Technik als Schulfach. Vol. 1. Düsseldorf, VDI-Verlag 1982
6. Huning, A.: Das Schaffen des Ingenieurs. Beiträge zu einer Philosophie der Technik. Düsseldorf 1974

7. Kultusminister Nordrhein-Westfalen (ed.): Richtlinien für die Gymnasiale Oberstufe in Nordrhein-Westfalen - Technik. Cologne 1981
8. Rekus, J.: Teaching Technology with a focus on moral education. In: M. Kussmann, H. Steffen (eds.). Current Topics of Technology Education in Europe. pp. 77-82. Düsseldorf 1991.
9. Ropohl, G.: Einleitung in die Systemtechnik. In: G. Ropohl (ed.). Systemtechnik - Grundlagen und Anwendung. pp. 1-74. München und Wien, Hanser Vlg. 1975
10. Steffen, H., Weiler, T.: Sociogramme of a complex system - Computerprogramm zur Förderung ganzheitlichen Denkens. In: M. Kussmann; H. Steffen (ed.). Current Topics of Technology Education in Europe. pp. 93-99. Düsseldorf 1991
11. Theuerkauf, W. E.: Technikunterricht und Berufsorientierung. Grundlagen für eine didaktische Konzeption. Bad Salzdethfurth 1983
12. Theuerkauf, W. E., Weiner, A.: Qualifikationsanforderungen an Absolventen der allgemeinbildenden Schule aufgrund veränderter betrieblicher Strukturen. In: M. Kussmann, H. Steffen (ed.). Current Topics of Technology Education in Europe. pp. 17-27. Düsseldorf 1991
13. Theuerkauf, W. E., Weiner, A.: Key qualifications as a reference system in technical training. In: Technological Literacy, Competence and Innovation in Human Resource Development. Proceedings of the First International Conference on Technology Education. April 25-30, 1992. pp. 408-415. Weimar, 1992
14. Theuerkauf, W. E. Weiner, A.: The German Dual System of Vocational Education and the Implications for the Human Resource Development in the United States. Paper read to the annual congress of the American Vocational Association. Nashville, TN 1993
15. Wagener, W., Schmidt, K. H.: Technik, Energieumsatz in technischen Systemen. Bochum 1981
16. Wiener, N.: Kybernetik. Regelung und Nachrichtenübertragung in Lebewesen und in der Maschine. Düsseldorf 1963
17. Wolffgramm, H.: Allgemeine Technologie, Elemente, Strukturen und Gesetzmäßigkeiten technologischer Systeme. Leipzig 1978

Perspectives and Concepts in the Swedish National Curriculum for Technology – and a Modest Proposal for a General Model That Will Explain It All – Hopefully...

Thomas Ginner
Linköping University
Department of Technology and Social Change
S-58189 Linköping, Sweden

Abstract. A new national curriculum for technology is just being implemented in the Swedish compulsory school system (grade 1 to 9, age 7-16). This paper presents some of the basic structures and concepts introduced in the new syllabus and a model to give the teachers some guidance when interpreting the text and planning their work. The paper has a limited purpose: to serve as a starting point for a discussion on content and methods in relation to the aims and how the subject could be structured.

Keywords. Curriculum, technology education, subject model

1 Background

After more than twelve years of investigations and testing in schools, Sweden got its first National Curriculum (NC) in 1962. It was a document that in a rather detailed manner prescribed what was expected to take place in our schools. Since then we have had three revisions, the latest one was published early 1994.

For many reasons it is not possible today to produce such documents as the NC of 1962. It was a detailed document that may have been appropriate at the time. But the effect of that type detailed steering document seems to have been rather marginal. The ideas behind the NC of 1994 reflect this experience.

Accordingly, the new curriculum is aiming at more liberty *and* responsibility for the local school and its teachers when it comes to choice and structuring of content and methods. Hopefully this new system will increase the teachers professionality. (The whole idea is threatened by a new marking system with national criteria for every subject and marking level.)

So instead of presenting detailed regulations, the new subject curricula will have two different types of goal. There should be open-ended aims, describing in what direction the students should strive – and the distances they will cover along that path will vary. The other type of goal sets the minimum level every school must reach. Every student should – in principle – reach a certain minimum level evaluated in grade five and nine. If not it is supposed to function as a signal to the school to improve the teaching. So it is not an evaluation assessing the individual in the marking system.

In the NC of 1962 technology was a subject within the vocational programmes and was soon labeled as a subject for the less able boys. The system was changed in 1969 and the differentiated programmes for the age range 12-15 was abandoned as well as the vocational training. However, the students still had a certain number of optional subjects, among them technology, to choose between, but the problem of segregation was not solved.

In the national curriculum of 1980 this system was changed and developed. It gave more freedom to schools and teachers and technology was introduced as a compulsory subject at all levels in the Swedish schools. This was of course a new thing. But ten years later we can establish the fact that the compulsory technology subject has not been a success. There has hardly been any technology teaching in primary and lower secondary. One reason is the lack of a structured curriculum for technology. With the NC of 1994 this situation has changed.

2 The Revised Curriculum for Technology

When constructing the curriculum there were three traditions and misconceptions of what technology education is about that must be 'killed' or altered.

- Technology taught as applied or extended science
- Technology as a very narrow, vocational and strict practical subject
- 'Unstructured walks' in daily life technology (which often was synonymous with jumping from the toilet to the bike to the washing-machine etc.)

Technology must be looked upon both as a knowledge area in its own right and at the same time interrelated to other subjects. The figure below is one way to describe it. To deal just with the technological aspects is far too narrow. The way technology is understood in the new curriculum is illustrated by the figure 1.

Since we were not allowed (and did not want) to prescribe content and methods we had to find some basic perspectives by which the essence of technology and its relation to its context could be understood. In the final text from the Ministry for Education these perspectives were formulated as follows:

Historical and International Perspectives
Technological innovations and their social consequences from a historical and international perspective must be considered important aspects in technology education.

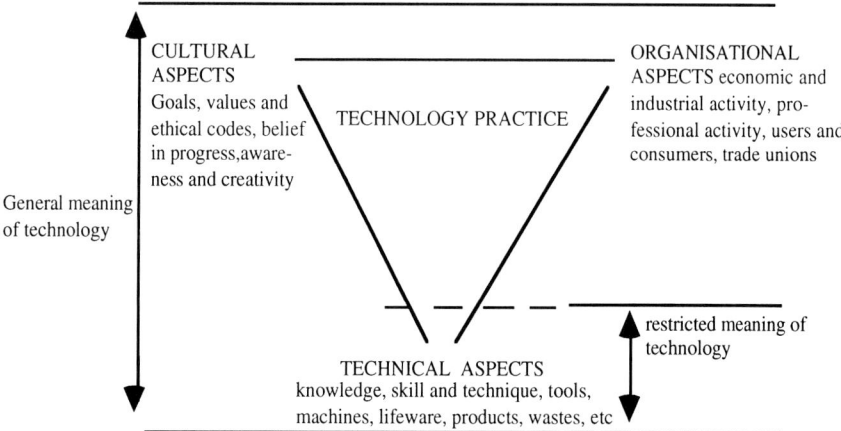

Fig. 1 Pacey, A. (1983) The Culture of Technology. In Layton, D. Values in Design and Technology (ch 3). In Budgett-Meakin, C. (ed) (1992). Make the Future Work.

The Social and Environmental Aspects of Technology
Technological education must illustrate the consequences for individuals, for society and for the environment of particular technological innovations. In doing, opportunities will arise to exemplify and discuss a range of ethical, economic, political and social issues.

The Purposes of Technology
Technology clearly serves a variety of purposes and functions. Four important purposes and functions can be identified: transformation, storage, transport and control (including the notions of steering and regulation).

Technology as a System of Components
Much of modern technology consists of more or less advanced components which, collectively, can form more or less complex systems. By examining individual technical components and their role in the complete artefact or system, students can gain valuable insights into the particular character of technological activity and the conditions required for it to flourish.

Technology as Construction and Application
Technological education must be concerned with the problem-solving character of technology, expressed in systematic development from identifying a need or a problem, to constructing and evaluating a solution. To understand and become familiar with technology and technological principles, students must engage in a variety of techniques and with a range of technological problems.

These are the perspectives given in the curriculum. The text above is partly rewritten and shortened, but the different aspects are presented in the same order and roughly with the same explanations.

3 A Way to Re-structure the Aspects

Since the curriculum was introduced we have had comments from both primary and secondary teachers. They sometimes find it difficult to see how they should deal with the perspectives above. Even if the text in the new curriculum is slightly more extensive than above it still does not give much assistance. The teachers are left with very little guidance and good examples. At least so far. During the initial work with the curriculum the following aspects emerged - but was never adopted by the Ministry:

1. The component-system perspective
Technology is composed of more or less advanced components, which in their turn can form more or less complex systems. By working with the chain *component-tool/machine-system*, studying separate technical objects, how these are linked together and form systems the students can gain important insights into the special character and conditions of technology.

2. Construction and operation
To be able to understand and become familiar with technology and technical principles, the student must both practically and theoretically test how different technologies and technical solutions are constructed and how they operate.

3. The tasks of technology
What technical function does the technology being studied have? Here we can identify at least four important functions :

- transforming
- storing
- transporting
- controlling, steering and regulating.

4. The interplay between technology and man - driving forces and consequences
In order to understand the role of technology and its significance, we must look at the relationship between human needs and technical development. What are the driving forces behind technology, how can change in a specific area of technology be explained? What are the consequences or effects for the individual, society and the environment of a certain technology?

5. The historical-development perspective
Studying how man has developed and used various technologies, what the driving forces and the effects have been for the individual, society and the environment, is a way of increasing our understanding of today's technology, its modes of application and basic principles. It also gives us an understanding of more important historical changes and processes and helps to gain insight and develop plans of action for the future.

It is not that difficult to see the parallels with the final aspects in the official curriculum and the ones described above. But the latter can be linked to each other in a way that could help us to see the relationship between technology, man and nature. It also has proved to be to some assistance for teachers when planning their work. The links could be described with the following figure.

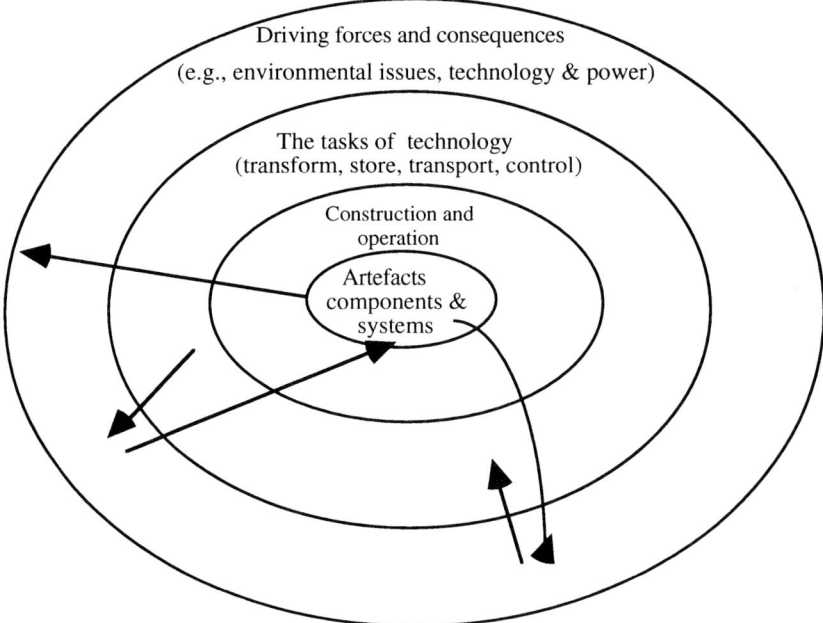

Fig. 2

The model is a way to try to explain how the different aspects on technology interrelates. The technological artefact is in the centre. It could be studied as a component but also on an aggregated level with several components forming a system.

In the next 'interface' we study the function or work carried out in the middle. The design of the construction and the way it works is in focus here.

In the third 'interface' we will find the different functions technology has; transform, store, transport and control/steer.

And it is the consequences in this sphere that shape the conditions in the fourth one. That is where we find both the driving forces behind technological development and the consequences of the use of technology.

The model has two aims: to help people see the structure of the links between technology and the context and to give the teacher an educational map. They do not have to start in the middle and walk outwards or begin in the outer sphere and walk towards the middle. The teacher and the students can take off anywhere, the map is there to inform them where they are.

A New Curriculum for Technology Education in Schleswig-Holstein

Gerd Höpken
College of Education Flensburg
Institute for Technology and Its Didactics
Fruerlunder Straße 37, D-24943 Flensburg, Germany

Abstract. The existing curriculum for technology education in Schleswig-Holstein - valid since 1986 - is structured in fields of action: working and production, transportation and traffic, building and built environment, supply and waste management, information and communication. In 1992 a revision of all curricula began. Guidelines for this revision were:

1. New curricula should lead to a discussion on core problems (e.g., basic values of living together in peace in different cultures; preservation of the elements of life; future change of economic, technological and social conditions; equal status of women and men; the right of all human beings to form their political, cultural and economic conditions).
2. New curricula should secure a common basic education.
3. New curricula should enable the interweaving of learning experiences.
4. New curricula should enable co-operation across the bounds of school subjects.
5. New curricula should facilitate lessons.

These guidelines led to a completely new structure. Now the first units of the new technology education curriculum are finished.

Keywords. German school system, technology curriculum, technology education

1 The German Educational System

After World War II, the Federal Republic of Germany was established after the model of the United States of America. Figure 1 shows the federal states ('Länder') and their population figures. In the German Federal Republic the federal states are independent in educational matters.

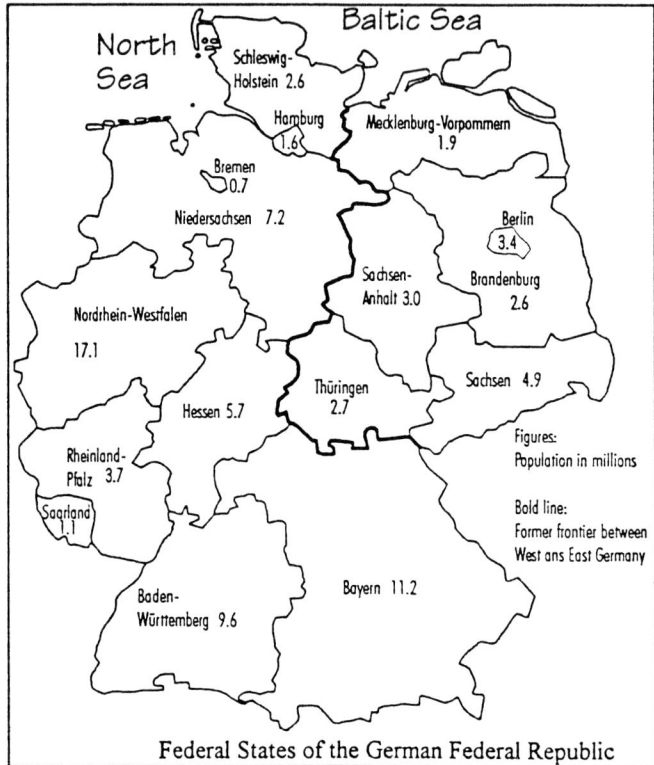

Fig. 1 Federal States of the German Federal Republic

In spite of the independence, the educational system in the federal states is nearly the same (Figure 2). Although in all states there are comprehensive schools, the normal system consists of a primary school and three branches of secondary schools: Hauptschule (general secondary school), Realschule and Gymnasium (high school). In some states the first two years of secondary school form a school on its own, the Orientierungsstufe.

The school leaving certificate of the Hauptschule entitles the student to enter into an apprenticeship. The same does that of the Realschule, under certain conditions the student can go to the Gymnasium. The leaving certificate of a Gymnasium entitles the student to go to a college or university. Today roughly one third of the pupils attend each of the branches Hauptschule, Realschule and Gymnasium. Only a few pupils attend comprehensive schools.

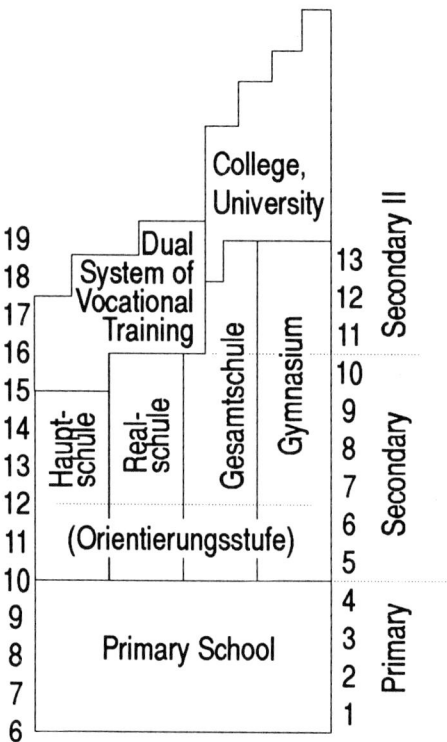

Fig. 2 The German Educational System

In most of the federal states technology education is taught only in Hauptschule and Realschule (Figure 3).

2 Revision of Curricula in Schleswig-Holstein

To meet the common problems of today's schools, the Schleswig-Holstein Ministry of Education decided to revise all curricula of generally educating schools. The revision began with an evening opening ceremony in the Castle of Kiel on December 6, 1991. The inaugural address was held by secretary of state Bodo Richter: Educational policy for the world of tomorrow. Only two more speeches were held that evening. Hans Heinrich Driftmann - chair of the Schleswig-Holstein commission of Educational Policy of the Society of Employers' Associations - declared "Demands of the economy for a future oriented educational policy". Martin Baethge - professor of sociology of Göttingen University- described "Educational expectations and demands for qualifications - educational sociological cornerstones for tomorrow's society".

40 Gerd Höpken

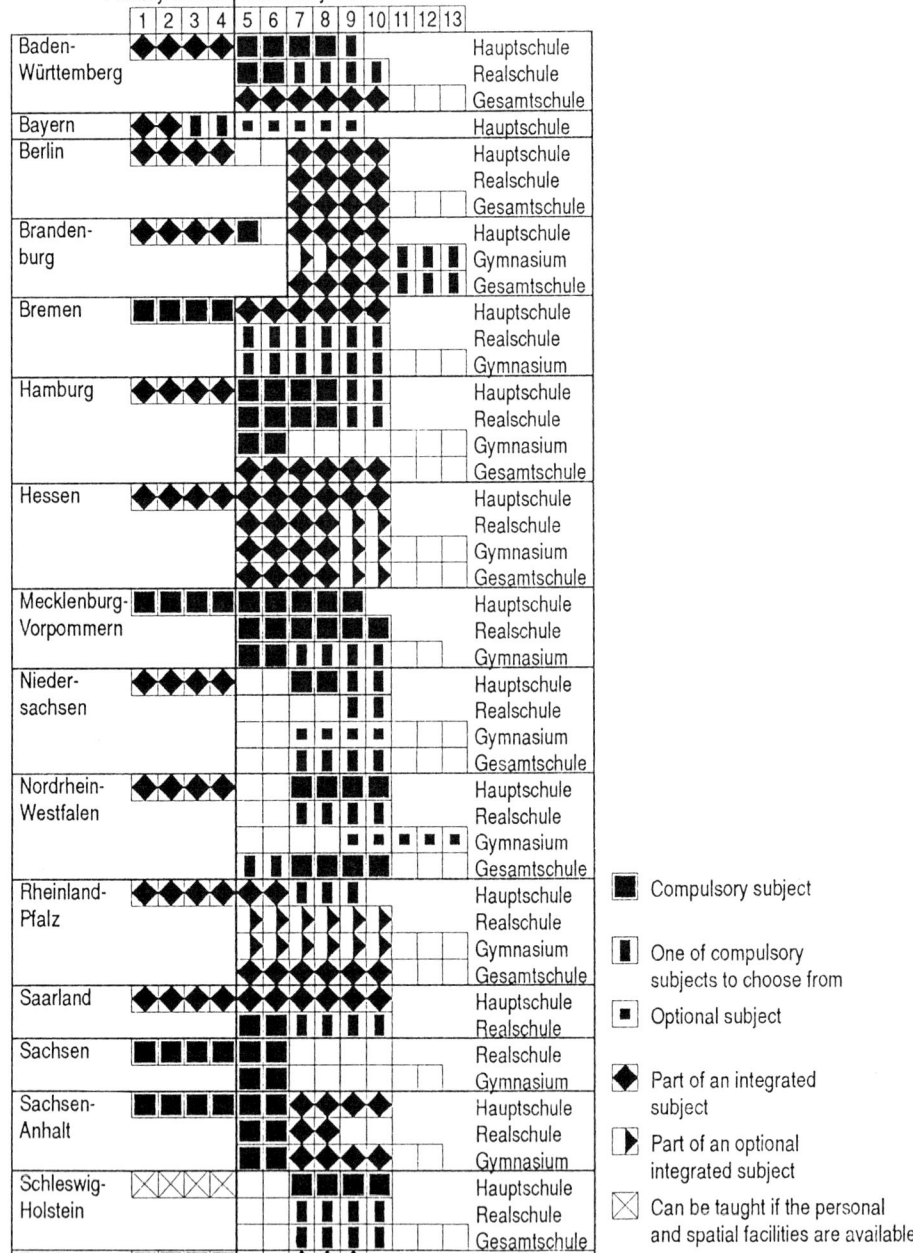

Fig. 3 Technology Education in German Schools

The event continued on the next day at Kiel University. Here the frame conception for the revision of the curricula was explained by Klaus Karpen, head of the department "Basic Problems of Educational Policy" at the Ministry of Education. He referred to the Prime Minister's policy statement on internal school reform. The main targets were:

– opening lessons in respect of subject matter, methods, and organization
– discussing new tasks and challenges
– expanding cross curricular and cross school branches co-operation
– democratising school life

The revision of the curricula was to take place in the Sekundarstufe I (junior secondary schools, forms 6 - 10 of Hauptschule, Realschule, Gymnasium, and Gesamtschule). The basic ideas are:

1. *The new curricula should come to terms with core problems*
Today's views on life and educational conceptions are hardly agreed on any longer. The conception of orientation toward science is no longer sufficient as the only reference point for teaching and learning. Nevertheless curricular work needs a common base. Such a base could be formed by working on core problems. Present core problems are:

– Basic values of human living together, in particular peace, human rights and living together in a world with different cultures, social systems, peoples, and nations as an individual and global task.
– Preservation of the natural fundamentals of life, of one's own health and other people's well-being.
– The future change in economic, technological, and social conditions of life and their impact on shaping conditions of life.
– Equal status for women and men, boys and girls, in family, vocation and society.
– The right of all human beings to organize their own political, cultural, and economic conditions of life, their participation and joint responsibility in all areas of life.

2. *The new curricula should secure a common basic education*
The explosion of knowledge and accelerated change hamper a joint participation of all people in culture, economy and society. That is the reason why cross school branches common basic education is necessary. An education in which the student can develop factual competence, judgement, ability to act and communicate. The target is to enable the student to participate in common assignments in school, vocation, and society.

3. *The new curricula should help to connect the experiences of life*
While in former times for most of the students school was the only important place of learning besides neighbourhood, today's students live in multiple worlds and correlation of learning.

4. *The new curricula should strengthen cross curriculum cooperation*
Subjects remain the most important way in school, to provide a variety of information for teaching and learning. The academic disciplines and their didactic studies therefore remain the basic frame of orientation for curricular work. The contents of the curriculum must have - in addition to the subject purposes - a discernible relation to educational targets and have to come to terms with the core problems. That is the reason why the work of the commission for each school subject is accompanied by permanent co-ordination of related school subjects.

5. *The new curricula should facilitate teaching*
Today great demands from everywhere are made on schools. A place for new subject matters can only be found when former ones are discarded. The different school branches have to shape a distinct profile. Learning happens in different ways and individually. Curricula must make room for free learning, internal differentiation, and individualized learning. Teachers need more free room for creativity.

6. *The new curricula should strengthen profile and cooperation of school subject, school stages and school branches*
The new curricula ought to show what has to be done in the lessons in a clear and intelligible way. But they do not substitute for the planning of lessons or academic articles.There will be curricula for each subject in each school branch, but they have to be well co-ordinated in order to secure a common basic education. The curricula will get a new structure: After an introduction explaining the position of the subject within the legal educational task the minimum requirements on graduated common basic education will have to be named. Not till then will follow school branch specific structured learning programme. The curricula must be made in a way that a sensible layman can understand them and can estimate their intentions and organization.

In the last part of his speech Karpen explains the principles of the process. For each subject in each school stage a commission is made up. Besides the teachers this commission will consist of representatives of parents and students, and experts from academic disciplines and teacher education. Parts of the new curricula will be tested. The results of the testing will be fed back to the commissions.

In the documentation of the curricula revision, there is an organization scheme, which shows the participants and their role in the process (Figure 4).

In the further course of the conference, papers were presented referring to the core problems. The headings of the papers are:

Core problem 1: Basic values of human living together, especially peace, human rights and living together in a world with different cultures, social systems, peoples, and nations to be an individual and global task.

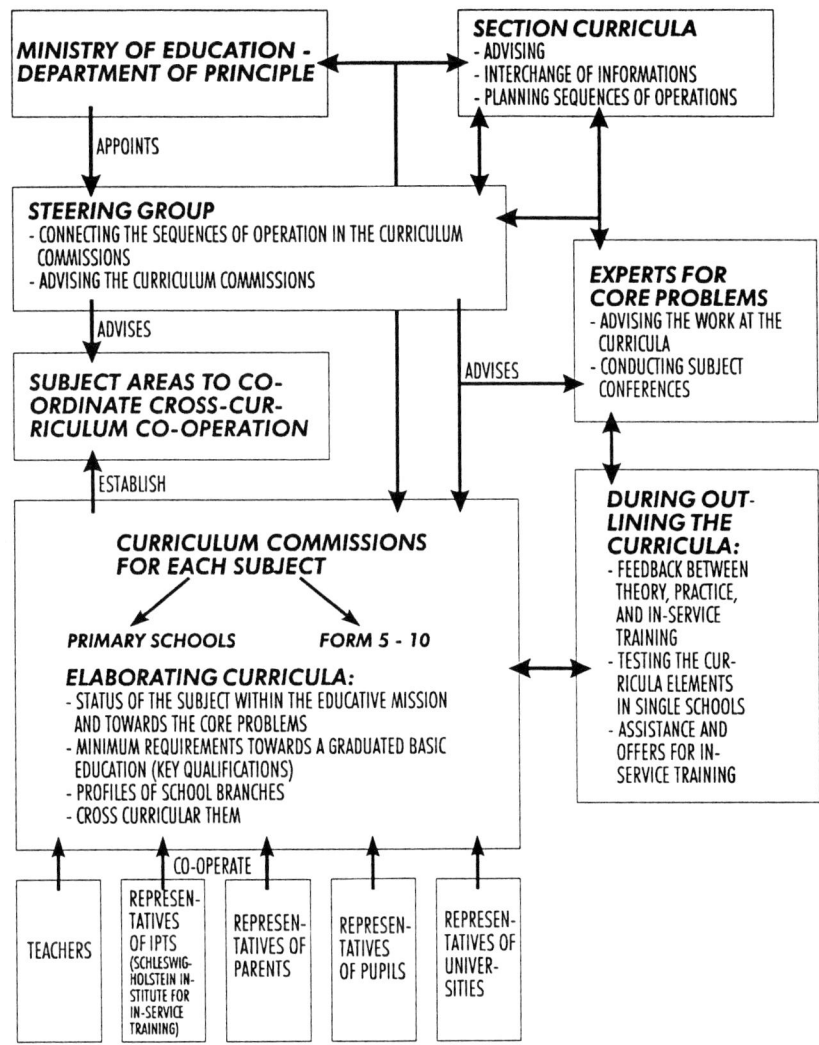

Organisation Scheme of Curricula Revision in Schleswig-Holstein

Fig. 4 Organization Scheme of Curricula Revision in Schleswig-Holstein

1. Basic values of human living together and their meaning for school
2. Principles of education in regard of peace
3. Peace as an aspect of subjects, the example of sciences
4. Projects in peace education
5. Overdevelopment or surviving
6. Learning in a world which is a part of 'One World'

Core problem 2: Preservation of the natural fundamentals of life, of personal and other people's well-being.

1. Ecological education - shaping the man-nature relation as a guideline for pedagogy
2. From subject lessons to cross curricular ecological education
3. The future task of ecological education as a duty for all school subjects
4. Learning in a healthy way is more than learning health

Core problem 3: The future change in economic, technological, and social conditions of life and their impact on shaping conditions of life.

1. Landmarks for curricula revision Schleswig-Holstein
2. Records of working group 1 (economics)
3. Records of working group 2 (social)
4. Records of working group 3 (new media)

Core problem 4: Equal status for women and men, boys and girls within the family, vocation and society.

1. Research outcomes relating to the gender issue - consequences for lessons and pre- and in-service training of teachers
2. Records of the working group 'research outcomes relating to the gender issue'
3. Language aspects of inequality and equality of gender
4. Unequal conditions for girls/women and boys/men in the areas of science education and vocational orientation

Core problem 5: The right of all human beings to organize their own political, cultural, and economic conditions of life, their participation and joint responsibility in all areas of life.

1. Social change and participation in society, economy, and politics
2. Intercultural education - charge and chance
3. Intercultural district schools - bilingual learning

3 Revising the Technology Curriculum

The commission for technology education consists of 6 teachers (two each from Hauptschule, Realschule and Gesamtschule), a representative of parents, and a member of Flensburg University.

The main problem in the first session of the commission was the fact that teachers stuck to the old curriculum, which was very clearly formulated. Especially the structure following the fields of action was a help for teachers in structuring the lessons. The core problems were rather understood as a sort of political indoctrination than a help to make a curriculum. But with the discussion going into the depth of the problems, members of the commission began to recognize that the core problems were realistic problems for the students in the present and in future. In this phase of the work some of the members of the commission quit co-operation and new members were selected. Selecting new members happened in agreement with the existing commission, candidates were invited for 'test' sessions.

First the commission formulated key qualifications for technology education:

- acquiring factual knowledge about technology
- acquiring ability to organize production in craftsman's or industrial ways
- acquiring safe ways of acting
- knowing fundamentals of order at a working place and in the workshop
- using appropriate ways of working with materials, tools, machines, and devices
- making objects in a accurate way
- acquiring sensitivity towards sparing resources, using energy, avoiding waste, and recycling
- preparing ability to use technology in safe and appropriate ways
- bringing creativity into the solution of problems
- acquiring technological sensitivity and awareness of problems/learning transferable strategies
- acquiring ability to comprehend complex connected situations
- acquiring ability to evaluate technology
- acquiring decision competence in the area of pre-vocational orientation
- acquiring ability to experience
- acquiring independence
- discovering and perceiving one's own capacities
- acquiring ability to work in teams and to accept each other in different situations

To show the connections of key qualifications with the core problems rough objectives were defined for each field of action:

- Work and production
- Transportation and traffic
- Supply and waste management
- Information and communication
- Construction and built environment

Structure of the teaching units. During the elaborating of the first units the way in which to integrate the reference to core problems was discussed. The following structure evolved:

Specification: The specification includes an orientation towards a target.

Reference to the core problems: The connections of the subject matter with the core problems are set forth.

Contribution of the subject matter to basic education: Here the multiple aspect view of the subject becomes clear. The commission developed a structure diagram to explain the fields of tension of the subject matter. This diagram also shows connecting points to cross curricular teaching.

Reference to key qualifications/teaching intentions: The listed intentions connected with the subject matter are compulsory. They are oriented towards the key qualification.

Remarks referring to the subject matter / pedagogical remarks, cross curricular aspects: The subject matters should be understood as offers. The intentions can be worked out with other examples.

Once this preparatory work was done, the teaching units were worked out. The following overview shows them (Figure 5).

Sociotechnical field of action				
Work and production	Responsibility of man working with raw material in craftsman like production. Basic course: Communication in technology. 7. - 9. form	Development and employment of machines change place of work and vocation. Interdependence of man and machine in production. 7. - 9. form	Industrial production of article for daily use and its impact on conditions of life. 8./9. form	
Transportation and traffic	Bicycle technology and appropriate use of means of transport. 7. form	Car technology and its interactions with man and environment. 9. form	Technology conceptions for environment conserving means of transport. 10. form/project	People develop technology (e.g., air craft engineering) and use it in different ways. 8. - 10. form

Construction built environment	Former and present ways of constructing bridges- basic principles of static, selecting materials, impacts on man and environment. 7. form	People protect and secure themselves - safety systems of yesterday, today, and tomorrow. 7. - 8. form	Dwelling in changing times -ecologically beneficial, human building and living together. 8. - 10. form/ project	
Supply and waste management	Wrapping is a burden for environment - disposing and planning wrappings, avoiding refuse through abolishing, recycling relieves environment. 7. -9. form	Supplying and disposing garbage of a household under technological, ecological and economic aspects. 8. - 9- form	Using energy efficiently and sustainable energies in households. 9./10. form	Man as consumer - discriminating dealing with the supply of technical articles - analysing, testing, and purchasing products. 9. - 10 form
Information and communication	Basic electrical circuits and safety education. Basic course· Soldering. 7. form	Impact of automation technology on man, working place and vocation. From hand control to computers 7. - 10 form	Interchange of information, development and impacts. From the drum to wireless telephones. 8. - 10. form	bold: compulsory subject matters

Fig. 5 The 1994 Technology Curriculum, Overview

The comparison of both surveys shows that the new curriculum is less rigid. It gives more freedom to the teacher to teach the matters in different ages. From the comparison is not visible that the new curriculum contains much more text to ensure the reference to the core problems. The large amount of text was the reason why the dissemination has come out to be more difficult than that of the old curriculum. Teachers had to be introduced to the curriculum in in-service training courses. To facilitate spreading the curriculum, the commission developed graphic representations to show the intentions.

The first presentation (Figure 6) shows the connections of the fields of action with the levels of technology education.

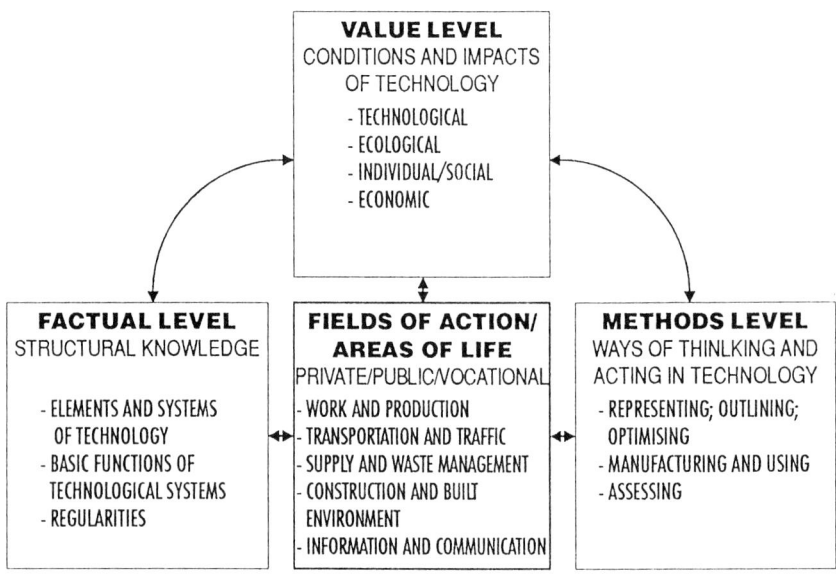

Fig. 6 Didactical Structure of Technology Education

The second representation shows the interdependence of technological aspects with aspects affected by technology. Such a graphic representation was made for each teaching unit. With these illustrations, the new curriculum grew again.

4 Main Differences Between Previous and New Curriculum

The previous curriculum had preliminaries explaining the targets of technology education, comments on methods and media of teaching, remarks on planning and carrying out lessons, and a survey of the curriculum. This part comes up to 18 pages.

In the new curriculum, this part contains a lot more of information (e.g., core problems) and amounts to 30 pages. In the previous curriculum, the representation of the curriculum units takes 72 pages. The structure of these units is very simple: objectives, subject matter, and advice for teaching are presented in parallel. The structure of the new curriculum is a little more complicated. As shown above, every unit has a preliminary note of its own. It contains a rather problem-oriented title than a short heading. The connection with the core problems is discussed, and references to the key qualifications are shown. After these preliminary notes advice for the subject matter proposals are given. The teacher is always free to select other matters to meet the teaching intentions! Parallel to these proposals, teaching hints are offered, combined with suggestions for cross curricular work. Though the size of the units differ, they at least cover 16 pages each. (Not all are finished yet.) So all 17 curriculum units will sum up to approx. 300 pages in addition to the 30 pages preliminary notes.

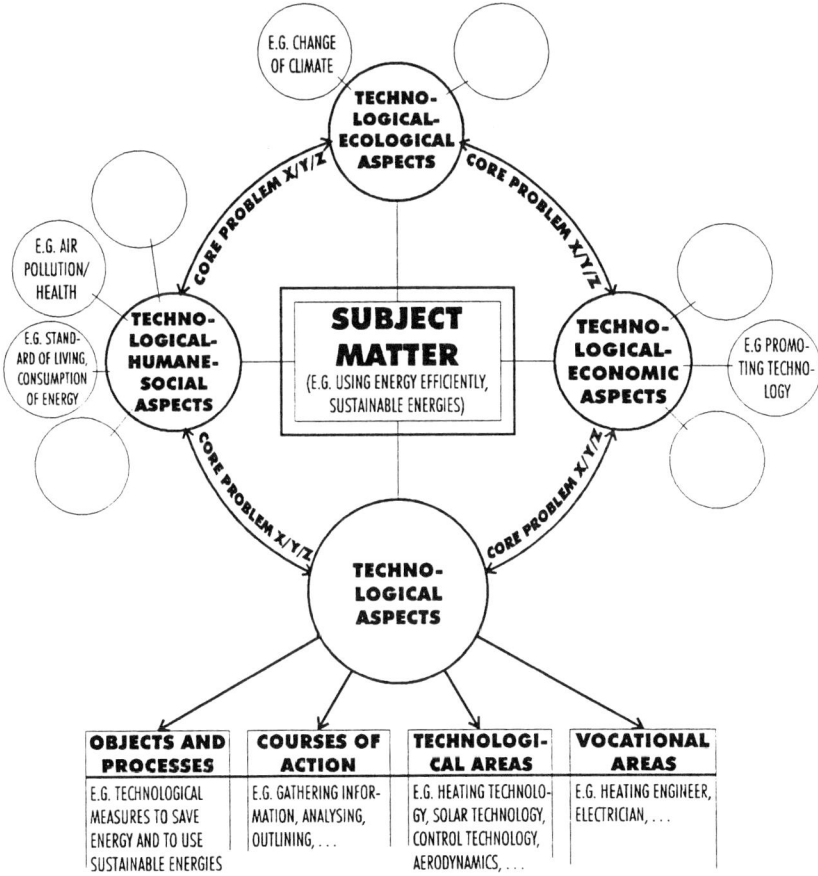

Fig. 7 Interdependence of Subject Matter

5 Conclusion

The new curriculum tries to show clearly the impacts and interdependencies of technology. With this intention they tend to become very complex. Teachers have problems translating them into lessons. They will need several courses of in-service training. Attempts to integrate the new curriculum in pre-service training seem to be more promising.

The new curriculum will not be valid until the commissions for all school subject have finished their work. So the Schleswig-Holstein government can take some time to decide about the new curriculum.

References

1. Der Kultusminister des Landes Schleswig-Holstein (ed.): Lehrplan Realschule Technik. Kiel, Schmidt & Klaunig, 1986
2. Die Ministerin für Bildung, Wissenschaft, Kultur und Sport des Landes Schleswig-Holstein (ed.): Lehrplanrevision in Schleswig-Holstein. Kiel, 1992
3. Schulte, H., Wolffgramm, H., Hartmann, E., Hein, Chr., Höpken, G.: Allgemeine technische Bildung - Technikunterricht. Stuttgart, Klett, 1991
4. Lehrplanentwurf Technik Sekundarstufe I (not yet published)

New Education Law, New Curriculum

Lajos Radics
Eszterhazy Karoly T.
Kepzö Föiskola
Szabadsag u. 2, H-3300 Eger, Hungary

Abstract. The subject called Technical Studies deals with our technical environment and with the application of technical knowledge in every day life. In Junior Schools Technical Studies is taught in either version A or B: the chief defect of the two conceptions.

Keywords. Curriculum, education law, technical studies

The environment surrounding the individual consists of social, technical and natural components. All three components determine the quality of life in the same way and the individual has to adapt himself to the circumstances and has to learn about it as well.

In our school system the subject called Technical Studies deals with our technical environment and with the application of technical knowledge in everyday life.

Technical Studies have been taught at elementary level for a hundred years. In the last century the public education law framed by Eötvös was passed, and elements of agriculture and industry were taught at elementary level.

In the 20th century the scope of technical knowledge widened and it became a dominant subject in the four-year higher elementary school. As far as Technical Studies are concerned, in the second half of this century the syllabus of Technical Studies in the 8-year elementary school system was based on the syllabus of Technical Studies in higher elementary schools.

In the last decades the name of this subject has changed, today it is called Technical Studies, and it is obvious that in elementary schools we need this subject and we will need it in the future because of the progressive technical development in Hungary and abroad. A pupil should find his way in his technical environment and that is why he should be familiar with it and should know a lot about it. He has to be familiar with the main areas of technical culture and he has to learn how labour, industry, agriculture and environment are interdependent and how technical knowledge can be used for and against technical development.

The aim of Technical Studies is to give fundamental knowledge about the operation and function of technical mechanisms because these mechanisms form a part of our environment, to reveal the possibilities and restraints of this subject. Some people like this subject, others reject it, depending on their temper, expertness and competence or incompetence. Nevertheless nobody can be unconcerned with it. Its appearance as a school subject is the result of social demand in Hungary. In Hungary Technical Studies as a subject was introduced in elementary schools in 1978, in universities in 1979, in colleges including our college in 1982.

University training is 5 years and after it the qualified teachers of Technical Studies start teaching the subject in the first and second forms of higher schools. College training takes 4 years and qualified teachers of Technical Studies teach the subject in junior schools (children between 10 and 14), where version A or B is taught. In teachers' training schools training takes 3 years and after it primary school teachers conduct practical classes in infant schools (children between 6 and 10). These practical classes prepare pupils for their further technical studies.

In junior schools Technical Studies is taught as a distinct subject, and pupils learn either version A or B of Technical Studies. The choice depends on the form of the settlement where the pupil lives. In urban and industrial areas the teaching material is based on the technical knowledge of industry. This form of Technical Studies is called version A.

In rural areas and villages pupils learn version B. In version B half of the teaching material is technical knowledge and the other half is based on agricultural knowledge. In a few elementary schools including the practicum schools of our college both versions are taught.

At our college the first curriculum of Technical Studies was worked out in 1981 but it was incomplete because teachers graduating from our college were able to teach version A. Version B was left out of consideration and the old curriculum had no reference to the structure of the Hungarian economy/ industry - agriculture/ and to the demand of society.

The Committee of Technical Studies wanted to change the old curriculum. It was improved by the Ministry of Education in 1984. A new subject, of agricultural studies, called Agrotechnics, was introduced into the new curriculum. We began teaching it in the third semester and students had a two-hour lecture and a two-hour practice a week.

The number of classes was not enough and we compiled the newest curriculum. It was drawn up by our department in 1988 and the Committee of Technical Studies did not take part in working it out. It was introduced on 1st September 1989. The quitting principles to the new curriculum were stated by the ministry which insisted on some regulations. For example the ministry determined the number of semesters: 8 semester for regular students and 10 semester for correspondence students. The total amount of lessons for two subjects is 2500 lessons. The number of lessons for one subject is 1250 lessons. The maximum amount of lessons is 29 lessons a week and this amount is divided into two parts - 20 lessons for the main subjects and 9 lessons for the other subjects. In respect of examinations there is a minimum of 3 examinations and a maximum of 6 examinations in a semester and there is a maximum of 2 complex examinations in a semester.

There are no other restrictions and regulations. Our department determined and chose the subjects we wanted to teach. We left the redundant subjects out of the new curriculum and we introduced new subjects to satisfy the needs of society. We had to have discussions about the new curriculum with the other related departments. Our curriculum is independent of what is taught at other universities and colleges.

And now I wish to give an outline of the curriculum worked out by our department. The curriculum was introduced in an ascending line. After finishing their studies students trained on the basis of the new curriculum have the alternative of defending their thesis or taking an oral state examination. The two alternatives will be equivalent.

The chief defect of the present system of teaching Technical Studies lies in the two conceptions. Children in urban areas learn Version A which does not contain agriculture, however, learning about agriculture would be very important for them. Children in rural areas learn Version B but it is not necessary for them, because they know quite a lot about agriculture and husbandry. They help their parents with breeding animals and working on the farm. We hope that this basically wrong concept will be changed in the future. The new education law will open up new dimensions.

We must mention some of the newest measures in the education law. These measures will specify our work at the department. This list is not complete:

- there will be no central government measures and planning, colleges and universities will lay down conditions under which students will be trained.
- this year number of applicants will not have to pass an entrance examination at some universities, and entrance examinations will be abolished in 1992.
- new youth and student organizations may offer a proposal and they will not have the right to intervene and the right of veto although they used to have these rights. The autonomy of the colleges and universities must be confirmed. Training must be confirmed. Training must be guaranteed at a high level.
- students may reduce the length of the training time and they may comply with the requirements of one or more semesters earlier than they are expected to.
- at state colleges and universities training will not be free of charge. Students will have to pay a school-fee (20000 Ft/semester).
- universities and colleges will confer honorary titles of assistant professor, professor, lecturer on specialists from industry and agriculture. These specialists may lecture at colleges and universities.
- as well as state schools, colleges and universities, private institutions, schools and religious colleges and universities will be established.

At last I want to mention that at our college we want to introduce university training at a few departments and later on university training will be extended to other departments as well.

New departments were established at our college. For example German and French departments, departments of American Studies, departments of philosophy and social science.

Some departments of our college moved to the one-time educational centre of the communist party and new college buildings are under construction. The name of our college has changed. The parliament passed a resolution. It declares that the name of our college has been Károly Eszterházy Teachers' Training College since 1st July 1990.

Research in Technology Education: Some Insights

George Shield
University of Sunderland
School of Education
Hammerton Hall, Gray Road, Sunderland SR2 8JB, United Kingdom

Abstract. The tradition of curriculum development in technology education has been based upon the implementation of educational ideals, some of which are fuelled by political/economic concerns and others based on learning theories or philosophical movements. This approach, however, whilst valuable, is not in itself sufficient. It is essential that such discussion is informed by the results of research into what is possible. Research, into technology education however, is at present sparse and is also influenced unduly by what is 'measurable'. As this approach leads to the neglect of whole areas of interest it needs to be enhanced by utilising techniques which are employed by other academic traditions such as historians, sociologists and ethnographers. The paper looks briefly at one such approach. The establishment of a resource base of grounded research is essential for the development of technology education.

Keywords. Curriculum development, technology, technology education, research methodology

1 Curriculum Development

Much curriculum development in technology education has originated from philosophical discussion which is informed by ideological, political, economic and other related considerations (Eggleston, 1992; McCormick, 1992). Such activity leads to the production of models and descriptions of the subject, and how these thoughts should inform what happens in our schools, both in terms of subject content and in its pedagogy. This philosophical debate is valuable. It focuses our minds on educational ideals, and it also requires us to question our practice. It makes us examine traditional methods, and encourages us to face up to the changing world.

This approach normally consists of proposals provided by the 'great and the good'. These are then thrown open for critical comment. Revisions may or may not be made following consultation and the proposals are then implemented (figure 1).

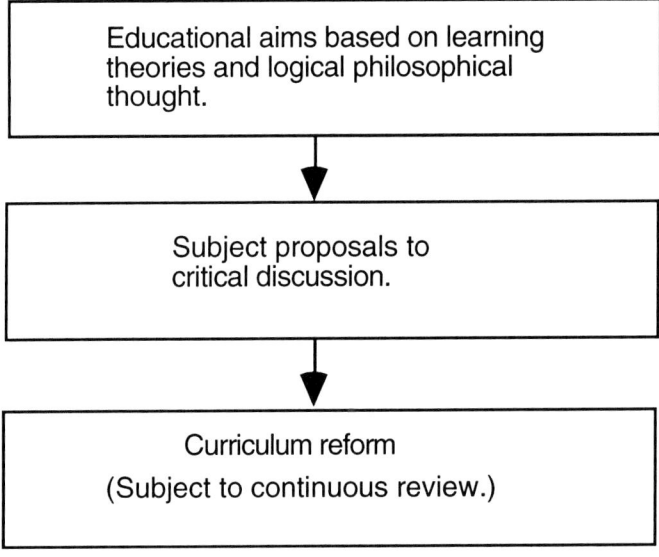

Fig. 1

The danger of such an approach is that views which are developed from these deliberations are then promoted as workable teaching and learning strategies. They become established truths and are accepted by many 'as gospel' because they have emerged from 'think tanks' which are either blessed with the power of the law (as in the United Kingdom) or they have authority vested in them by virtue of the credentials of established experts in the field.

This approach can lead to curriculum 'reform' which has been instituted with only limited empirical backing, without meaningful discussions with the people at the 'chalk face' and in some cases with only a minimal degree of trialing.

There are at least two dangers with the approach I have outlined. First of all there is the tendency in the consultation period to only receive views from established 'experts' and vested interests whose major (sole?) concern is to fight and defend their corner.

The second danger is the almost total disregard paid to what is actually happening in the majority of our schools. By this I mean not just one or two schools which are particularly expert and are used as show cases or by quoting the opinions of those who say *what they believe* is happening in their schools

Most teachers are doing a satisfactory and sound job. Let us cull from their practice the better aspects of their work and promulgate these across the wider national and international fields. It is important that we base curricular reform on what can be delivered. Education is like politics, it is the art of the possible. If you cannot take the people with you, through lack of will, understanding or skills and knowledge, the reforms will fail.

The difficulties which result from the current approach to curriculum reform has led me to add my voice to the increasing number of colleagues who have recognised this problem and to make a plea for the emphasis on developments within technology education to be placed on research based on empirical evidence rather than on philosophical considerations, valuable though these are (de Vries, 1992; Jenkins, 1992; McCormick, 1993).

One way forward must be to revise our approach to a more logical (and dare I say it, technological!) process and include the results of research into the decision making process.

Figure 2 outlines an approach which goes some way in meeting these shortcomings. To ensure a balanced input in the curriculum planning operation block B must have as much importance as block 'A' on the initial proposals. If 'B' does not exist, and it is very limited at the moment, curriculum reform must be postponed. If it is not postponed until an adequate research base is available the reforms will fail, resources will be squandered, your teaching staff will be alienated and your children will be betrayed.

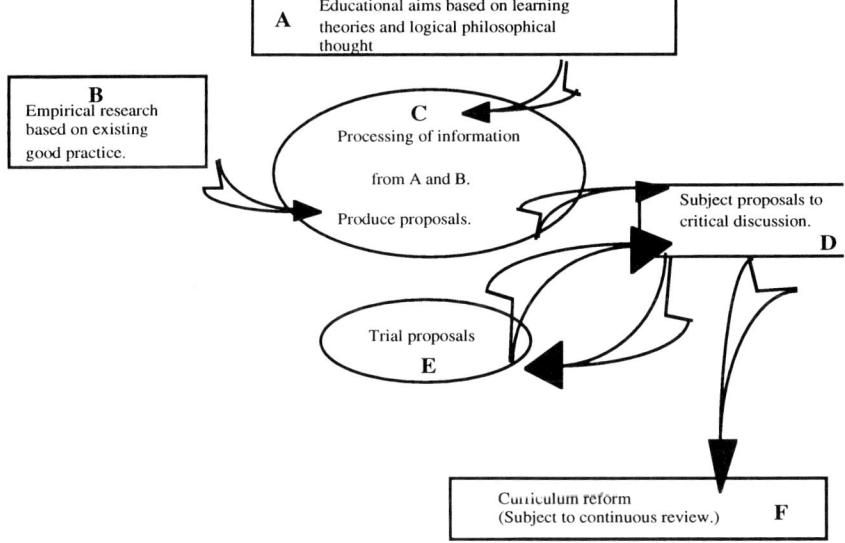

Fig. 2

A major problem in applying this approach however, is the dearth of published research which will be of assistance to the curriculum innovator.

The picture is not completely bleak, for example the Assessment of Performance Unit in the United Kingdom has in recent years published a substantial piece of work (Kimbell et al., 1991) which provides insights into the field of assessment in technology and the PATT research (Mottier et al., 1991) into perceptions of technology education are extremely valuable but in the context of a major subject area there is much still to be addressed.

In an analysis of the papers presented at Weimar in 1992 (Blandow and Dyrenfurth, 1992) I classified, and recorded the number of papers presented under six headings. The results of this survey are shown here:

1. The philosophy of technology education together with papers critically reviewing accepted views. — 39%
2. Papers describing how technology is delivered in various parts of the world. — 24%
3. Papers describing curriculum innovations. — 17%
4. Papers describing technological innovations. — 3%
5. Papers describing quantitative research. — 9%
6. Papers presenting qualitative research — 7%

Whilst this exercise is not meant to be a definitive piece of research it is indicative of the general trend. It can be readily seen that only something like 16% of the papers presented were based on empirical research. If this figure was trebled it would still not be out of place as a proportion of what I would consider desirable. I suspect that the conference is not unusual in the field of technology education and a similar picture could be obtained at most conferences in this subject area.

I do not want to leave the discussion at this point because I may be accused of dealing with only half of the problem and of doing exactly what I am complaining about, i.e., talking about talking. The argument may be advanced with an illustration of some research methods and examples placed in the context of current debate.

2 Context

The current trend in technology education has been towards emphasising the pre-eminence of the 'process' of technology over the acquisition of facts and skills other than those of a problem solving nature (HMI, 1987; N.C.C., 1990; Savage and Sterry, 1990; APU, 1993). These general problem solving skills are said to be transferable across boundaries and once acquired can be used in many different situations. This philosophy has resulted in a system which has elevated strategies which lead to a perceived ability in problem solving to a higher plane than technological understanding and which has also led to a dearth of those cognitive concepts more commonly associated with the subject content of 'technology'.

The logic of this argument is difficult to refute. What is technology in terms of subject content? Is electronics more important than food technology? Is biotechnology less important than the study of mechanics and structures? How does it all function without an understanding of economic imperatives? Within the finite resources of any education system, decisions have to be made over what to teach and what to exclude from the curriculum. Over what content to include within a subject and what to leave out. What is meant by 'basic' and what is the educational value of each item of technical 'knowledge'? If the subject could be re-packaged into some form of 'process' not only satisfying the sincerely held beliefs of a large body of educationalists as well as placating the competing vested interests of a range of 'subject bodies' and others, the whole problem could be solved.

Unfortunately this proposition that problem solving skills can be readily transferred to novel situations is receiving increasing criticism from an ever growing body of opinion, which whilst recognising the value of 'process' in education, is beginning to question the more extravagant claims which are made (Millar, 1988; Ormell, 1992; Hennessey et al, 1993).

If some form of empirical research had been carried out and the results included in considerations before the implementation of wholesale curriculum reforms in the United Kingdom, the advantages of a process methodology may have been gained without the harmful side effects, together with the consequent bad publicity, which have become apparent in the educational system of England and Wales (Smithers and Robinson, 1992).

3 The Research Process

What do I mean by 'research'? Anderson defines educational research as:

> "Research in education is a disciplined attempt to address questions or solve problems through the collection and analysis of primary data for the purpose of description, explanation, generalization and prediction."

(Anderson, 1990, p. 4)

Whilst Bassey offers this description:

> "Research entails systematic, critical and self critical enquiry which aims to contribute to the advancement of knowledge."

(Bassey, 1990, p. 35)

The definition of what counts as 'research' may not appear contentious. To some the use of terms such as 'investigation', 'inquiry' and 'study' are interchangeable. And to others research is something more technical and rigorous (and consequently only carried out by 'experts'). I would not like to enter this debate here but simply state that I regard that the key words are 'systematic' 'rigorous' and the 'dissemination of the knowledge gained', together with the inclusion of what Hammersley (1992) describes as the concept of 'relevance'.

Most technology teachers and university lecturers in this subject are numerate and tend to look towards a classical or 'scientific' approach to research. This gives comfort to the researcher in that it not only follows a well established technique but it often also gives security in the numbers that are created (Smith, 1975; Stenhouse, 1981).

This approach to research is usually determined by the availability of data which can be readily translated into numerical form and create the substance of valid conclusions. It can however, lead to a situation in which research is restricted to that which can only be measured. In other words the research methodology is restricting the scope and range of topics to be investigated. This classical approach has been criticized as being out of touch with the real world of education (Eisner 1985; Parlett, 1982) as it does not answer such questions as how or why particular events happen.

In technology education the search for data which can form the basis of informed comment is complicated by the nature of the learning process which takes place in the technology lesson. The range of concepts covered is extensive and the learning activity itself is based predominantly on a range of practical activities.

In practice the situations which are being examined are frequently 'one off'. The 'scientific approach' will often fail to recognize and take into account the unexpected and unpredicted occurrences which take place in the classroom. The tendency to disregard those areas of education which can not be validated by measured means can result in the neglect of meaningful events which may have a bearing on the educational process. The further qualification that the human attributes which are perceived as being measurable, such as tests of achievement, attitude ratings and levels of motivation, can have a range of interpretation which also leads to disquiet (Hopkins, 1990).

The basis of research into the process of teaching lies in an analysis of the relationship between the teacher and the taught. This analysis relies upon the subjective interpretation of what occurs in the learning environment and this interaction is frequently so complex that it can not be reduced to numerical data which is meaningful (Kincheloe, 1991).

Whilst therefore research based on a quantitative approach may in certain cases be of value and even necessary it cannot be said to be sufficient. Researchers in technology education must continue to expand their portfolio of instruments and utilize the much broader methodologies which are used by colleagues in the fields of history, sociology and ethnography (and indeed those devised by themselves) to complement those with which they are familiar.

The qualitative approach presents a broader, more embracing understanding of the situation under investigation (Rist, 1981).

4 Research Methodology

A major initial difficulty lies in defining or identifying the research question or establishing a hypothesis. The complexity of the learning environment frequently masks the inter-relationship between the many distinct but complementary factors involved, such as the environment, management, of the learning experience and curriculum content and pedagogy as well as the personality and intellectual attributes of teacher and taught.

One way forward is to realize that initially the questions which will enable the researcher to identify 'good' practice can not be identified i.e., the hypothesis, which is the foundation of a classical approach, cannot be formulated in advance, and strategies must be developed to aid the initial questioning which takes place. This approach can be illustrated in figure 3.

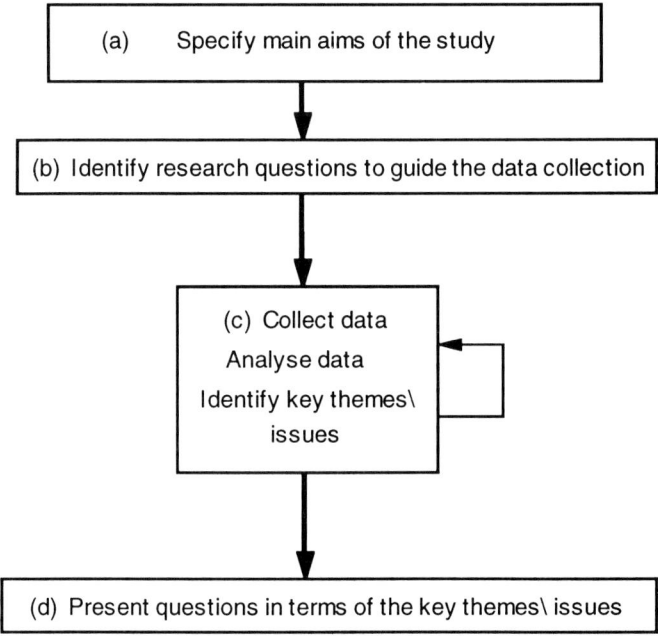

Fig. 3

In this diagram it can be seen that the research may have a theme which will provide a focus (a). For example the researcher may have a general interest in process methodology but will be unable to formulate a precise hypothesis. This interest may then be translated into questions which are of a general nature (b), such as "How does technology fit into the school structure?" "How do the teachers conduct their classes?" "How do the children learn?" In 'c' the collected data is then scrutinised to see if patterns of behaviour, or particular issues emerge. The results of this analysis (d) forms both the specific questions and provides a structure for insights into the practice of that teacher or institution.

The collection of relevant data to inform the research can come from a range of instruments with the choice as to the most appropriate being determined by the professional insights of the researcher and the nature of the investigation.

The basic tools of the work include interviewing, observation of the teacher in action, the use of a diary to record any occurrences which may have a bearing on the work of the school, available documentation and the scrutiny of other readily available sources of information. One of the basic problems is that of establishing the realities of the situation. A common difficulty is that as the work deals with humans there is nearly always a hidden agenda. The true 'facts' are difficult to identify and clarify through for example a questionnaire or structured interview. The tendency to produce the 'correct' answer or the response which pleases the researcher is strong. Each aspect of the work demands time to explore and possibly reveal the meanings behind responses.

The identification of relevant data can only be carried out by the 'connoisseur' (Eisner, 1985) and is achieved through constant review of the collection of data. In the example shown here the initial range of topics is constructed from data which has emerged from various sources such as informal discussions with teachers, conference papers, and journal articles following the use of the strategy outlined in figure 3.

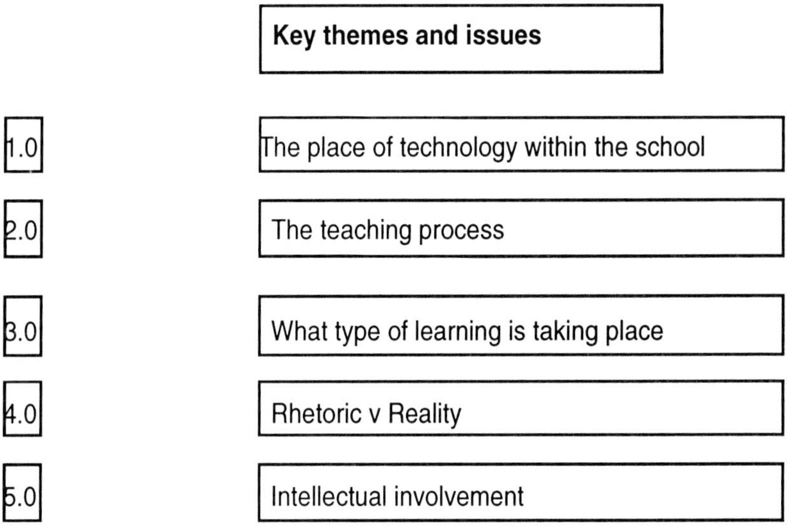

Key themes and issues

1.0	The place of technology within the school
2.0	The teaching process
3.0	What type of learning is taking place
4.0	Rhetoric v Reality
5.0	Intellectual involvement

Fig. 4

If this list is then broken down and subdivided into topics which are important for the subject a chart something like the one shown here (figure 5) emerges. Again these topics 'emerge' from a range of data. The data which informs this collection is fluid and is constantly being amended in the light of new insights which are being gained.

The headings for the classification in the example shown are not fixed, neither are they in order of priority. They are topics which appear to be significant in the bank of information which has been collected. This significance could for example lie in the regularity with which a particular topic occurs, or even the fact that it was very important in one school but not mentioned elsewhere. Why? Also it will is fluid and is constantly being amended in the light of new insights which are being gained.

The headings for the classification in the example shown are not fixed, neither are they in order of priority. They are topics which appear to be significant in the bank of information which has been collected. This significance could for example lie in the regularity with which a particular topic occurs, or even the fact that it was very important in one school but not mentioned elsewhere. Why? Also it will be seen that some of the data can be classified under more than one heading. (Tesch, 1990).

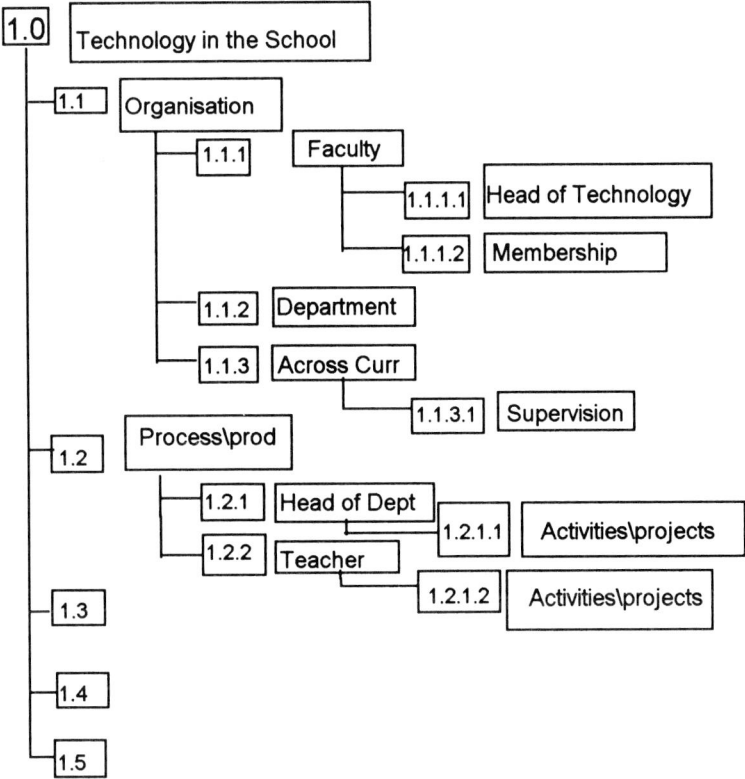

Fig. 5

When this initial categorization has taken place the evidence can then be extended and then interrogated to inform conclusions which will help the decision making process. This evidence could look like the following:

D 1.1 (*From department handbook*) The Technology area consists of independent departments representing the traditional areas of CDT, H.E., Art, Business Education and Information Technology. The work of these departments, for the purpose of the National Curriculum, is co-ordinated by the head of CDT who has this management responsibility delegated to him by the Head Teacher.

D 1.1 *(Interview with head of dept)* This approach is designed to retain the autonomy of the school's traditional subject areas which are recognized to have knowledge bases which are distinct but which are also seen to have elements, particularly in terms of methodology, in common. The majority of these common elements have been identified to meet the requirements of the National Curriculum.

Wh 1.1.1 *(From field notes)* Teacher T4 is the head of faculty. He had also entered teaching as a mature student having been working in an accounts department for a number of years. His initial teacher training was as a specialist craft teacher and all of his subsequent expertise has been acquired 'in-service'.

LF 1.1.1 *(From interview with head of department)* The academic organization of the school is based on departments. In Technology the contributing subject areas work together with the aid of an organising committee which is chaired by the head of each department in turn. The requirement for chairing this committee was 'written in' to the contract of all the heads of department when they were recently re-graded. This arrangement does not meet with the approval of all the staff as it appears to leave too much to the inclination of individual departments and responsibility is thought to be easily avoided.

> G.> How about your technology co-ordinator - is he or she...?
> N.> We don't have one. We are ... all the heads' of
> department are on C's and ... I was one of the first C's
> appointed in the school and then H.E. art, business studies,
> and IT were all on B's-but they were all upgraded, and as part
> of the upgrading it was written in that they were in charge
> of co-ordinating technology for one year.

D 1.1.1.1 *(From field notes)* The first session I observed was taught by the head of CDT department who is also the co-ordinator of technology.

Wh 1.1.2 *(From interview with class teacher)* Other points which emerged from this interview included the difficulties in reconciling the range of expertise required by the National Curriculum with expertise available. Whilst the 'carousel' system was thought to have advantages from this point of view, it was realized that a drawback was the difficulty in ensuring progression. In an ideal situation it was thought that a centralized facility may be of help in delivering the 'integrated' approach required.

M1.1.2 *(From school brochure)* The faculty of Technology includes the departments of CDT and Home Economics. Art is not part of this organization being seen to be part of an arts faculty but also as having a considerable part to play in its own right.

What could be noted from this, and much more data along similar lines, was that whilst the 'official' line of the sample of schools that I visited during this research was that the schools were based on faculties and all had technology co-ordinators. They were in fact functioning as departments and finding it extremely difficult to implement the National Curriculum along recommended lines (NCC, 1990). This information may not appear from a straightforward analysis of a questionnaire.

If I move from this example to one perhaps of more interest internationally, that of the practice of teaching, I have used different techniques.

In some cases I have analysed the movement of the teacher around the workshop to discover, the number and type of interactions which took place between the teacher and the taught and then plotted these movements on a chart. In the first chart the teacher is working with group of thirteen year old children who are in a workshop constructing a toy which has to have 'movement' built into it. (The work is based on mechanisms and includes levers, cams and gears.)

The second teacher is working in a design office and the pupils are designing a package for biscuits. This teacher was also involved in teaching about developments (nets) of unfolded containers.

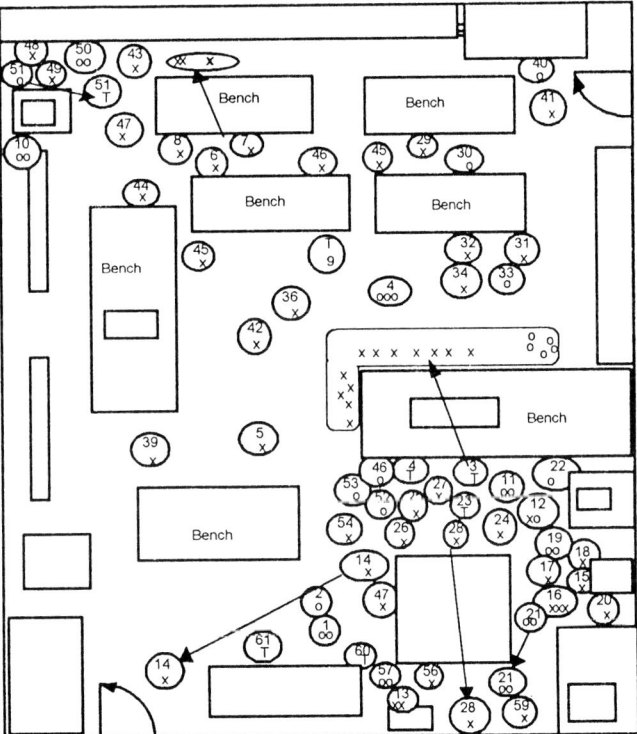

Fig. 6 O - girls, x - boys, T - whole group

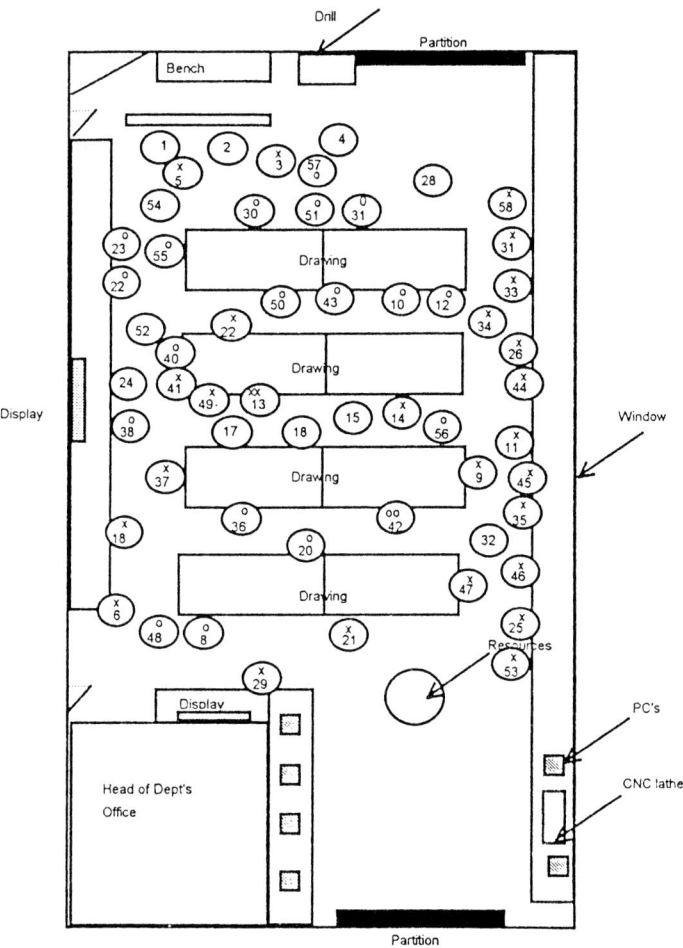

Fig. 7 o - girls, x - boys

From these two examples (from different schools) it can be seen that the teachers work extremely hard physically and intellectually. They are constantly moving around the room interacting with each child, in the first case 'on demand' with the second teacher more systematically but again as the lesson developed, on demand.
Due to individual nature of the work they are also having to deal with a considerable range of 'problems' and therefore the work is also intellectually demanding. The layout of the room likewise dictates both the teacher's movement and consequently his ability to interact with the whole range of children. (This study backs up the warnings expressed by Layton (1986) when he pointed out the demands made on teachers by project work.)

What is perhaps more interesting however is the nature of the interactions taking place, i.e., just what are the teacher and taught talking about?

To look at this I wired up the teachers with a micro tape recorder, for a whole lesson, and then analysed the recording. It soon became apparent that a considerable amount of time was spent dealing with comparatively mundane, though essential, tasks, e.g., pointing out where to find materials, and preparing materials on machines which the children were not equipped to use. This interchange between a teacher and his pupil is typical:

P   Sir where's my folder?
T   Everybody's work is in there.
P   Sir where's the numbers for the clock?
T   In here.
P   Paper
T   What colour?
P   What colour is there?
T   There'll be some green and some blue. Some red, some grey, some black.
P   Sir can I have some red?
T   Yes. Go down to my office - you know at the end of the corridor on the filing cabinet. O.K. green and blue on the filing cabinet and in room 89, which is in the corridor in the brown drawing cabinet- in the third drawer up from the bottom. Some large sheets of sugar paper that's where you'll find the red.
P   Sir where will I get ... for that.

In these following two examples however it can be seen that not all teachers interact the same way.

Mr John was more concerned with 'thinking' skills.

T   Right then Edward tell us how we got on with this..
P   ...
T   Do you think that's going to work? That's going to have to be a little bit wider. Do you know what ... do you know what perhaps we should do?... I'm not sure about that dovetail there. I'm not so sure that it should perhaps be a straight spigot going out. Either that or you're going to have to open up this space perhaps a little bit. next time you mark off ... mark off near the edge. Well try it ... There's nothing lost by just cutting one out and just trying it. If it doesn't work then modify it. So what system are you going to use?

He used this approach through most of the lesson. Constantly moving around advising on design principles, making techniques and very often, economy in the use of materials.

Mr Simon, however, from a different school, was far more concerned with getting the facts across I analysed his interactions in the following way:

(AS field notes) In a detailed analysis of a period of one hour during one of his lessons the following components of his teaching were identified:

Questions	
Open	Closed
28	52

During this session the children were engaged in individual work. It was interesting to note that the majority of the questions he asked were of a closed nature, these were used to elicit problems encountered by the children. He would then proffer advice or demonstrate some technique or process. The open ended questions were used to draw from the children their thinking on a particular topic. Again this was then used to extend the children's knowledge base.

This teacher was particularly prolific in giving information to the children and the type of advice and the number of times it was given during one lesson was noted.

		Information		
Process	Content	Facilitate	General	Admin. Instruction
6	60	35	6	14

If I then move on to a further, and perhaps the most important example, of how and what children learn I have used different techniques again. I realize, of course that we cannot get inside the child's head but we can observe and record their behaviour.

In this case I have placed a tape recorder near their work station whilst a group of children were designing a mechanism for a 'robotic' arm. (They were working on an adaptation of a bicycle brake mechanism.) This piece of data gave some interesting insights into group dynamics as well as the process they were going through. (As this conversation between the children was conducted in regional dialect there may be difficulty in translation!!!)

P1   Mine'll work won't it?
P2   Should do.
P1   Ya naa the bit that gan's like that and the bit that tak's the loop, and the wire gaans in and oot there. That'll be really tight an all.
P2   Small and tighter. Normally you pull the wire longer and ... where's the book?
P1   I think that'll get smaller ... but the wire'll get bigger.

LATER

T   How much was it?
P3   We'll measure the square right ? Then we'll know the distance we'll take for the square you put it in. You measure the distance what'll be when you put it upside down.

Here the children are problem solving by discussing designs amongst themselves, a self selected group. They have recognized the need to use reference material and are engaged in mathematical concepts (Are they 'internalising through doing'? Kimbell et al., 1991).

In another case I used a concept mapping technique to try to find out what the children had learnt from one teacher about mechanisms (Bougon, 1983; Elbaz et al., 1986). I spent some time explaining what a concept was and the purpose of the concept map before setting the group to work.

I classified the responses separately under boys and girls and divided the concepts under three headings:

1. The 'scientific/technological' concepts of 'mechanisms', i.e., responses which referred to 'levers' 'cams' 'linkages' etc. These could be said to reflect the 'content' or cognitive learning which took place during the lesson.
2. Concepts which mentioned objects such as machines, i.e., cars, drills and computers. These could be said to reflect a lay persons view of mechanisms.
3. Concepts such as 'energy' and 'efficiency'. These could be said to indicate a deeper understanding of the more abstract facets of the topic.

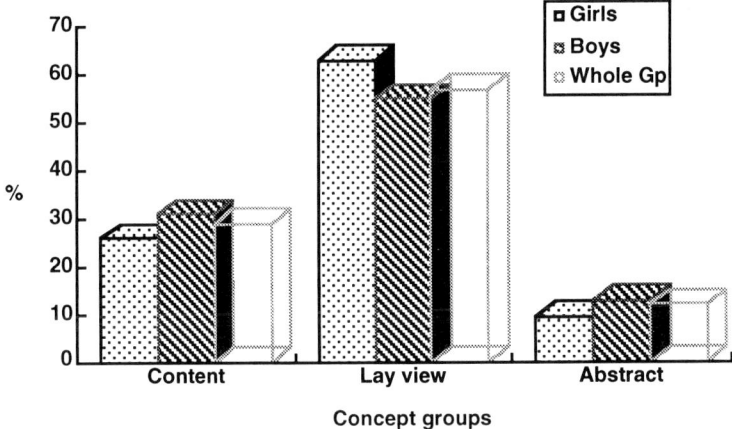

Fig. 8

When the results of this experiment were reviewed it was unsurprising to see that the largest response was in the 'lowest level', or the area I have termed the 'lay level', with 63% of the girls responses and 55% of those emanating from the boys recorded here. The overall figure was 57%. This result would suggest that the children had a large residual background knowledge of 'technology' which could have been acquired through learning experiences outwith the technology class as well as part of a structured learning programme. This knowledge could well (and probably did) arise from experiences which were not part of a formal learning activity.

In another school I endeavoured to find out what type of learning was taking place by analysing the internal test papers which a teacher had set his pupils. At this school the importance of subject knowledge and conceptual understanding was reinforced through the use of a formally structured and administered paper and pen test which was used to evaluate the knowledge gained and to supplement the subjective evaluation of the project itself.

The test paper included questions designed to test high order activities such as evaluation together with the recall of factual information. The *knowledge* base of the children was tested through *seventy per cent* of the questions with the remainder devoted to reasoning activities.

This highly factual approach to teaching can be seen at work in this example of a design brief which was set for the children in the same school:

Design Brief   Year 8 Design and Technology

A manufacturing firm has identified a market for electronic games which rely on the manual dexterity (Hand skill) of the players.
Design and make a prototype for a new game.

Specification

The game must:
1. Use a 9v battery
2. Use a light emitting diode (LED)
3. Use a resistor (330 ohms)
4. Use a buzzer
5. Be made from available materials.

School H. (Extract from design brief set for year 8 children)

The example shows a highly prescriptive approach to teaching a particular electronic circuit with a thin veneer of 'designing'. The children in effect ended up by 'designing' a switch.

5 Discussion

From work such as this small incomplete study a number of questions have emerged which may benefit from further reflection:

It begins to appear that whilst some schools say they offer an integrated faculty approach to technology, in practice they do not. They tend to work in smaller groupings using organizational structures actively discouraged by the Department for Education. Is this a more productive way to utilize teachers abilities than is officially recognized? Is it a better way of teaching technology despite learning theories suggesting otherwise?

Why do some highly effective teachers who say they are teaching 'process methodology' as their central concern, concentrate far more heavily on transmitting traditional 'technological facts' and skills than they recognize?

Is it possible to divert teachers from having to spend a great deal of their time in low level activities, which do not make the best use of their talents and which are also very demanding in terms of physical energy and intellectual effort into a more productive role? Is this question linked to my first two? What is the total picture of teacher activity during the design and make cycle of a project brief?

Here for example I have developed a provisional model of what this interaction may be like.

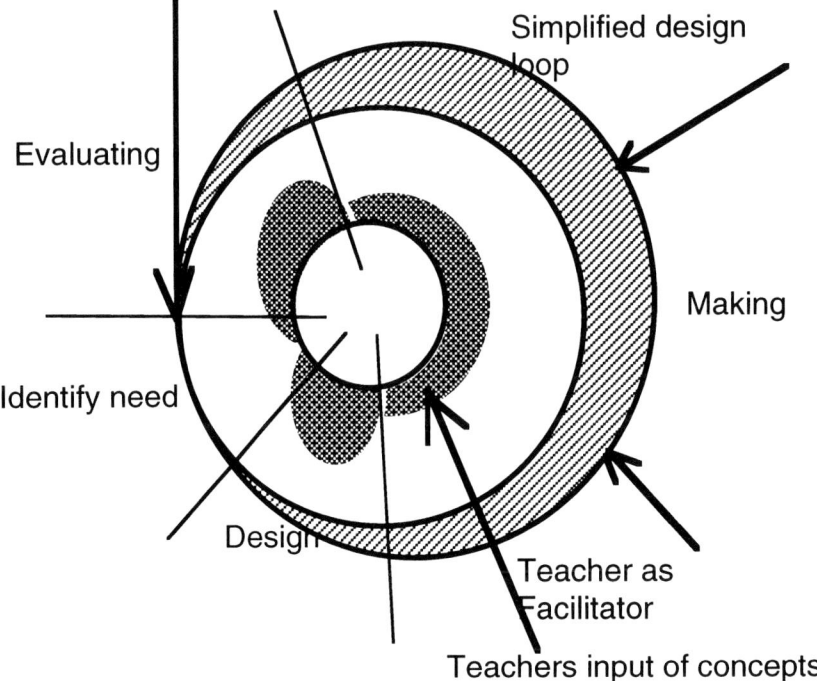

Fig. 9 The Teachers' activity during a design and make task

In this model I have assumed that the design and make task is a loop, similar to the many theoretical models of which you are aware (HMI 1987; APU, 1993). From my observations it appears that the teacher's input consists of two basic varieties, (there are other inputs) the first is to do with 'understanding' and can range across a very wide array of skills and knowledge, the second is to do with the no less important task of facilitating, or progressing, the work of the children. These two phases of activity peak at different stages in the technological process.

If we bear in mind that as each child is working as an individual and that the model given above represents the teacher's interaction with one child and that it could be repeated up to twenty plus times in a group of children, with each model slightly out of phase, it can be seen why the teachers in figures 6 and 7 are having to work very hard. It may also explain why they prefer to have an organizational structure which allows them to take as much control of the learning situation as possible and try to limit the unforeseen occurrences which occur when they are working in unfamiliar territory.

6 Validity of the Research

One of the most common criticisms levelled at research of this nature is the apparent lack of objectivity and validity in the findings obtained. This is a limitation which has to be recognized at the outset of the research and attempts made at all times to eliminate researcher bias and methodological shortcomings. Obviously all research is subjective in some degree and this is reflected in the questions asked and the conclusions reached. For example Scarth and Hammersley (1986) recognize the conflict between what the intentions of a teacher are in setting a task or carrying out a particular course of action and the researcher's interpretation of this action. In many cases, this can be overcome by a post activity discussion between the teacher and researcher, and also when written reports are submitted for checking, but it is recognized that this is not fool proof and differences in interpretation will occur. As well as this discussion with participants in the research, bias can be confronted through subjecting the resultant opinions to an examination by critical friends.

There should also be a thorough employment of a wide range of instruments. Field notes should be kept, and interviews taped. Some of the lessons can be taped to keep an accurate record of the teacher pupil interaction and photographs taken and kept as evidence of the environment (Dieckman, 1993). Other records, such as pupil work sheets and school documentation should also be available for scrutiny and to check for accuracy of interpretations (Stenhouse, 1980). It is in these terms that the validity of the research is recognized. The validity of the work is interpreted as "the correspondence of knowledge claims to the reality investigated" (Hammersley, 1992 p. 196)

Obviously single samples of such small scale research will not of itself be of great value, but a large collection will be. It is the only with the establishment of a bank of qualitative research to complete the picture provided by the large scale funded research projects that a valid picture of 'the possible' will emerge. Curriculum development will not only be substantially informed but we will all be in a much stronger position when explaining and defending the activities which take place under the banner of technology education.

References

1. Assessment of Performance Unit: Learning through Design and Technology. In: McCormick, R., Murphy, P., Harrison, M. (eds.) Teaching and Learning Technology. Wokingham. Addison-Wesley/Open University, 1993
2. Anderson, G.: Fundamentals of Educational Research. London, Falmer Press 1990
3. Bassey, M.: On the Nature of Research in Education (Part 1). Research Intelligence. No. 36. pp. 35-38, (1990)
4. Blandow, D., Dyrenfurth, M. (eds.): Technological Literacy, Competence and Innovation in Human Resource Development. Proceedings of INCOTE 92. Weimar 1992
5. Bougon, M. G.: Uncovering Cognitive Maps. The Self-Q Technique. In: Morgan, G. (ed.) Beyond Method: Strategies for Social Research, Newbury Park, CAL, Sage 1983
6. Dieckman, E. A.: A procedural check for researcher bias in an ethnographic report. Research in Education. Vol. 50 pp. 1- 4 (1993)
7. Eggleston, J.: The Politics of Technology Education. In: Blandow, D., Dyrenfurth, M. (eds.) Technological Literacy, Competence and Innovation in Human Resource Development. Proceedings of the First International Conference on Technology Education. Weimar 1992
8. Eisner, E. W.: The Art of Educational Evaluation. Lewes. Falmer 1985
9. Hammersley, M.: Some reflections on ethnography and validity. International Journal of Qualitative Studies in Education. Vol. 5 (3) pp. 195-203 (1992)
10. Elbaz, F., Hoz, R., Tomer, Y., Chayot, R., Mahler, S., Yeheskel, N.: The use of concept mapping in the study of teachers knowledge structures. In: Ben-Peretz, M., Bromme, R., Halkes, R. (eds.) Advances of Research on Teacher Thinking. Lisse, ISATT/Swets & Zeitlinger 1986
11. Hennessy, S., McCormick, R., Murphy, S.: The Myth of General Problem-Solving Capability: Design and Technology as an Example. Curriculum Journal 4 (1) pp. 74-89 (1993)
12. HMI: Craft Design and Technology 5-16. London, HMSO 1987
13. Hopkins, C. D., Antes, R. L.: Classroom Measurement and Evaluation. Itasca, Peacock 1990
14. Kimbell, R., Stables, K., Wheeler T., Wosniak, A., Kelly, V.: The Assessment of Performance in Design and Technology. London. SEAC\HMSO 1991
15. Jenkins, E. W.: Towards an Agenda for Research in Technology Education. In: Blandow, D., Dyrenfurth, M. (eds.) Technological Literacy, Competence and Innovation in Human Resource Development. Proceedings of the First International Conference on Technology Education. Weimar 1992
16. Kimbell, R., Stables, K., Wheeler, T., Wosniak, A., Kelly, V.: The Assessment of Performance in Design and Technology. London. SEAC\HMSO 1991

17. Kincheloe, J. L.: Teachers as Researchers: Qualitative Inquiry as a Path to Empowerment. London, Falmer 1991
18. Layton, D.: Innovators' dilemmas: recontextualising science and technology education. In: Layton, D. (ed.) Innovations in science and technology education. Vol. 1. Paris, UNESCO 1986
19. McCormick, R.: The Evolution of Current Practice in Technology Education. Part 1. The Journal of Epsilon Pi Tau. Vol. 18 (2) pp. 19-28 (1992)
20. McCormick, R.: The Evolution of Current Practice of Technology Education. Part 2. The Journal of Technology Studies. Vol. 19 (1). pp. 26-32 1993
21. Millar, R.: The pursuit of the impossible. Physics Education. Vol. 23 pp. 156-159 (1988)
22. Mottier, I., Raat, J. H., de Vries, M. J. (eds.): Technology Education and Industry. The proceedings of the Pupils Attitude Towards Technology Conference. No 5. Eindhoven. PATT foundation 1991
23. National Curriculum Council: Technology: Non-Statutory Guidance. Design and Technology Capability. York. NCC 1990
24. Ormell, C.: Is 'process' good for your health. Cambridge. Journal of Education. Vol. 22 (2). pp. 227- 242 (1992)
25. Parlett, M.: The New Evaluation. In: McCormick, R. (ed.) Calling Education to Account. Milton Keynes. Open University 1982
26. Savage, E., Sterry, L.: A Conceptual Framework for Technology Education. Reston, VA. International Technology Education Association. 1990
27. Scarth, J., Hammersley, M.: Some problems in assessing the closedness of classroom tasks. In: Hammersley, M. (ed.) Case Studies in Classroom Research. Milton Keynes. Open University Press 1986
28. HMI: Craft Design and Technology 5-16. London, HMSO 1987
29. Smith, H. W.: Strategies of Social Research: The Methodological Imagination. London. Prentice Hall 1975
30. Rist, R. C.: On the Utility of Ethnographic Research for the Policy Process. Urban Education. Vol. 15 (4) 1981
31. Smithers, A., Robinson, P.: Technology in The National Curriculum London. The Engineering Council 1992
32. Stenhouse, L.: The Study of Samples and the Study of Cases. British Educational Research Journal. Vol. 6 (1) pp. 1-6 (1980)
33. Stenhouse, L.: What counts as research. British Journal of Educational Studies. Vol. 29, June 1981.
34. Tesch, R.: Qualitative Research Analysis Types and Software Tools. London. Falmer 1990
35. de Vries, M. J.: Pupils' Attitudes Towards Technology. In: Blandow, D., Dyrenfurth, M. (eds.) Technological Literacy, Competence and Innovation in Human Resource Development. Proceedings of the First International Conference on Technology Education. Weimar 1992

Human Resource Development – Innovative and Integrative Thinking of Education for Life

Dietrich Blandow
WOCATE Executive Secretary
WOCATE Office Erfurt
Schlösserstr. 9, D-99084 Erfurt, Germany

Michael Dyrenfurth
University of Missouri-Columbia
Technology & Industry Education
105 London Hall, Columbia, MO 65211, USA

Abstract. It is internationaly acknowledged, that there is a close correlation between the educational standard of a society and theire oconomicaly growth, there living style and living standard. It is shown, that the educational standard includes technological literacy besides ather components. It is sated, that the summary of subjects like Physics, Mathematics, Geography, Languages, Arts etc. are not the personality itself. The keypoint of the personality is there needs oriented creativity, which can be developed best thrugh technololical subjects in contrasting perspectives as a part of the human technological interface.

Keywords. Human Resource Development (HRD), innovation, creativity, technological literacy

Hail to the skillful, cunning hand
Hail to the cultured mind
Contending for the world's command
Here let them be combined

1 What is Technological Innovation/Creativity and Why is it Important?

On Creativity, Problem Solving and Innovation

Creativity has been termed "The ultimate driving force in industry" by Donnelley citing Copola in the Journal of Technology Studies (pending, 1994) – similarly, creativity "a company's most valuable resource".

Donnelley citing Mok, identified "two forms of creativity: The form of the artist and the form of the designer" (pending, 1994).

"Problem solving is what you do when you don't know what to do."

He (Donnelley) also called abstract thinking "a prelude to creativity". Following this, he pointed out that creators/innovators follow Helmholtz' stages of creativity/invention: "preparation, incubation, inspiration and verification". Edwards, Holmes and de Graaff (1973) addressed the issue of the actual nature of innovation, discovery in their words, with the following:

"It is also one of the least penetrable, most elusive facets of science. Discovery remains a mysterious, creative act that can no more be replicated than an act of artistic imagination (p. 63)."

Perhaps this actual point of inspiration, i.e., creation or recognizing of solution, is analogous to the 'jump' in energy levels experienced by electron shifting from one orbital to another.

What triggers the 'aha' experience?

- pattern recognition
- crystallization
- alignments

Innovation

- seeing new routes over and/or around barriers
- envisioning alternative solutions
- using new/different tool in new or different ways

Examples: uses for animal skin

- common answer is shoes and other leather clothes
- unusual answer to hold animals together

Factors promoting innovation

- stress, pressure
- systematic log
- hybrid ideas
- human interaction

Why focus on TI/C

- Fewer people entering labor market (20-40%)
- Increased importance of technology to GNP
- Aging population
- Importance of economics to nations

With respect to this latter point, note that 70% of world trade is in manufactured goods (Manufacturing policy project, 8/92). Given this, the link between a nation's GNP and the technological literacy and innovative capability of its people is direct and positive. While it is likely that this has been the case for some time, given the technology and information explosion, the actual nature of the equation has shifted from an old through a new to a newer equations that each give progressively more weight to innovation.

Table 1 Equations

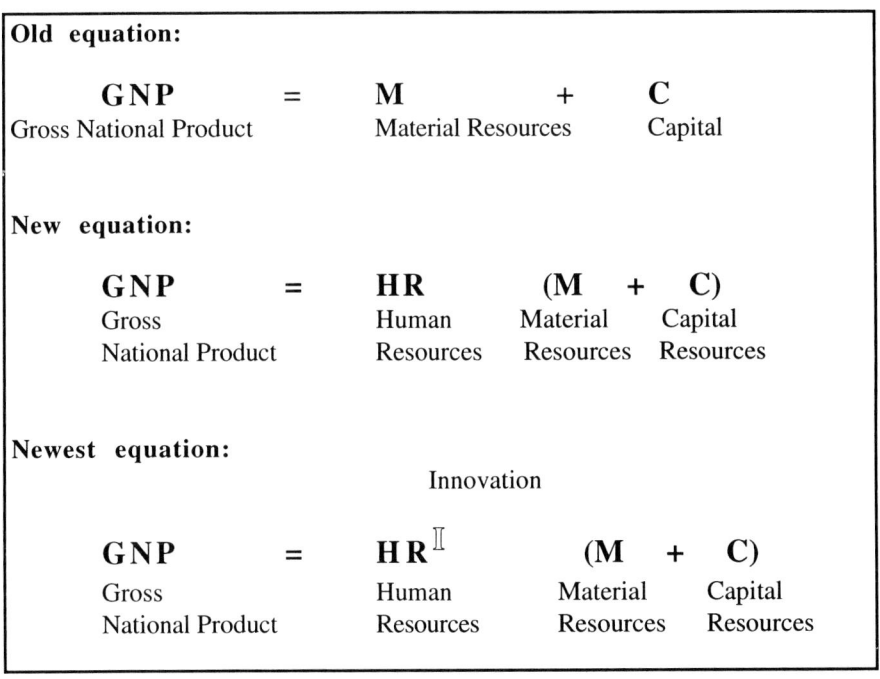

Furthermore, in addition to the obvious economically-based argument immediately preceding, note how the innovative potential of a person would also contribute significantly to his/her capability in each of the following outcomes resulting from the possession of technological literacy:

Table 2 Technological literacy outcomes

Technological Literacy Outcomes
• Knowledge of where to turn for help • Recognition of possibilities and limitations • Elimination of fear of, and instillation of respect for, technology • Triggers to imagination • Maximization of career possibilities • An ability to assimilation, interpret, and evaluate information

- Efficiency in the use of materials, energy and information
- Awareness of environmental implications
- Enhanced sensitivity to the implications for the human - technology interface
- Clarification of training needs
- A wider scope of reading/attentiveness

2 TI/C Models, Analogies and Metaphors

One of the ways of explaining TI/C is in looking to existing models, metaphors or analogies used to explain technological problem-solving and related procedures. Important among these are the:

- River model (Harrison)
- Taxonomic model (DeVore)
- Holistic model (Schmid)
- Didactic triangle of innovative thinking (Stonybrook)
- Step by step staircases model (Blandow/Dyrenfurth, 1991)
- Problem solving/Barrier model (Blandow/Dyrenfurth/Lutherdt, 1991)

3 What Do We Really Know about Technological Innovation/Creativity?

We know much more about technology than about technological innovation/creativity itself. Nevertheless, several key points seem to be likely true observations about TI/C.

Technological innovation/creativity is fueled by the energy of dissatisfaction, curiosity, fear, greed, desperation, in short all those forces that propel the human. Note that this does not exclude serendipity, coincidence, and dumb luck.

The TI/C process seems to involve the simultaneous search for discontinuities and patterns, i.e., every innovation emerges from contradictions and compromises. Furthermore, every innovation always occurs in context. There seems not to be the existence of innovation in the abstract. When considering the actual process by which innovation proceeds, it seems reasonable to conclude that innovation follows a process that is the opposite of that followed by evolution. One might even characterize innovation as a digital process as contrasted to evolution's analog process.

Innovation is not a singular event, instead it is a hierarchical chain of events. This conclusion was derived from an analysis of the necessary, desirable and sufficient conditions for innovation as shown in the table below:

Table 3 Analysis of conditions for innovation

Necessary	Desirable	Sufficient	Characteristic
Y	Y	N	• Understanding of technology systems involved
N	Y	N	• Understanding of technology backdrop/context
N	Y	N	• Understanding of individual/societal systems involved
N	Y	N	• Understanding of individual/societal backdrop/context
Y	Y	N	• Capability with technology involved
Y	Y	N	• Felt need/urgency
N	Y	N	• Perceived need
N	Y	N	• Awareness of projected futures/trends
Y	Y	N	• Ability to evaluate alternatives
Y	Y	N	• Awareness of solution criteria (ability to recognize solution)
Y	Y	N	• Available/accessible resources to act (time, money, materials, energy)
Y	Y	N	• Vision/Insight generating the solution

Note, no single characteristic carries the sufficient yes rating. To the authors this means that technology innovation/creativity is not a single event/action, e.g., the 'aha' or 'Eureka!' episode.

Instead, technology innovation/creativity is more likely a chain of events that individually (reductionalistically) don't comprise innovation is any way more than that the sum of the body's chemical substances comprise life.

An analysis of a series of interviews of highly creative people, soon to be published by Epsilon Pi Tau in their Journal, reveals several interesting commonalties across this very diverse set of people:

- Wide range of experience
- Multiple intelligences, typically running in parallel (distributed processing?)
- Enthusiasm for work
- Action along the interfaces of disciplines/fields
- Primary and secondary avenues to solutions, e.g., drawing and math; drawing and mechanics/modeling

To describe the context or locus of innovations, the authors' evolved a system that might be used to describe the context of the technology innovation/creativity process as locating the innovator's position along six or more key continua as shown below:

Table 4 Innovation Context Coordinate System

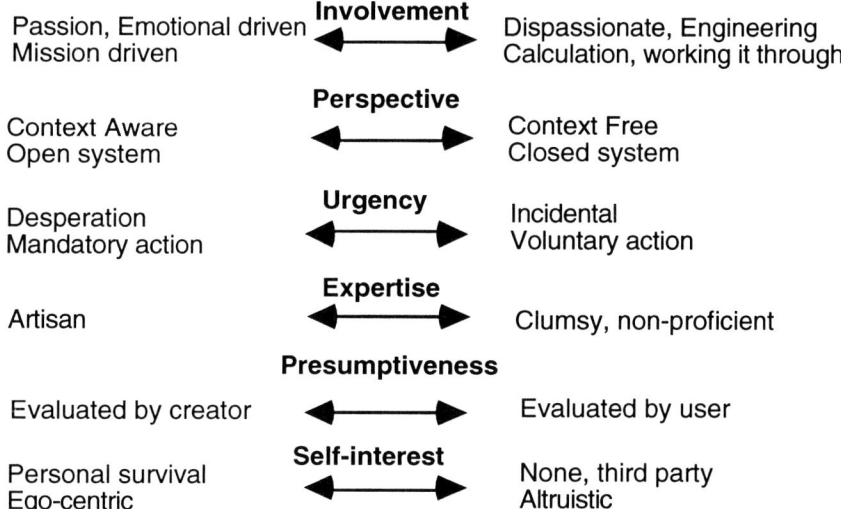

	Involvement	
Passion, Emotional driven Mission driven	⟷	Dispassionate, Engineering Calculation, working it through
	Perspective	
Context Aware Open system	⟷	Context Free Closed system
	Urgency	
Desperation Mandatory action	⟷	Incidental Voluntary action
	Expertise	
Artisan	⟷	Clumsy, non-proficient
	Presumptiveness	
Evaluated by creator	⟷	Evaluated by user
	Self-interest	
Personal survival Ego-centric	⟷	None, third party Altruistic

Finally, much can be learned from looking at the innovator, from their perspective, in terms of the resources actually available to them to fuel their innovation. To depict these, the authors used the following model:

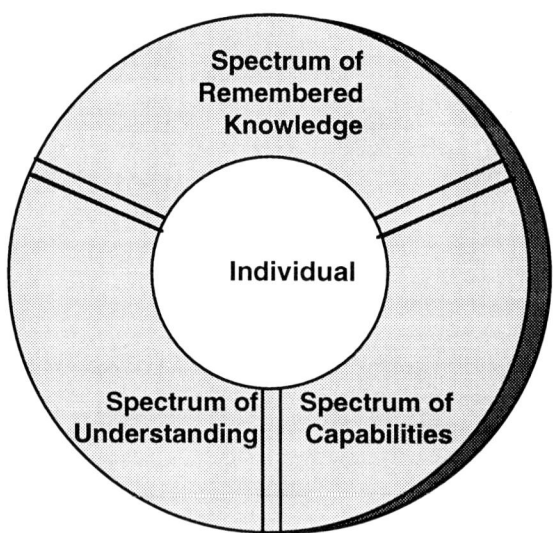

Fig. 1 Resources for TI/C

4 Needed Research Directions

To provide direction for readers as well as themselves, the authors then considered what was known about TI/C and projected key areas needing further research. These are listed below:

- Analysis of contemporary innovations' gestation process
- Human – technology interface (research into the human not technology side)
- Relative effectiveness of TI/C development strategies
- Investigation of context – innovation links
- ...

References

1. Bell, S., Sadlak, J.: Technology transfer in Canada. Research parks and centers of excellence. Higher education management, Vol. 4 (2), 1992 [EJ 453 108]
2. Coates, J. F.: Innovation in appropriate technology. Directions for policy research. Washington, DC National Science Foundation, 1980 [ED 217 887]
3. Dyrenfurth, M.: Technological literacy and innovation in industry. Presentation and paper delivered during lecture tour. Taipei, Taiwan 1993
4. Edwards, R., Holmes, B., de Graaff, J. V.: Relevant methods in comparative education. Hamburg, Germany UNESCO Institute for Education, 1973 [LA133. E3]
5. Liao, T.: Principles of engineering. Presentation to the NATO ARW on Advanced Educational Technology. Prague/Poprad 1993
6. Singh, R. R.: Education for the twenty-first century. Asia-Pacific perspectives. Regional symposium on qualities required of education today to meet the foreseeable demands of the twenty-first century. Bangkok, Thailand August 16-18, 1990 [ED 356 16]
7. Weese, J. A.: How NSF encourages industry--university partnerships. Engineering education, Vol. 75 (7), pp 646-669, (1985) [EJ 321 505]

MINDS 2000+: Innovations in Global Change as Part of Technology Education

James L. Barnes
Eastern Michigan University
Research Institute for Space Education
Ypsilanti, MI 48197, USA

Abstracts. MINDS 2000+ provides an innovative vehicle to addressing the national standards for mathematics, science and technology. It incorporates an interdisciplinary approach through hands-on/minds-on uses of technology. By using Internet and other technologies we can provide real world applications to the existing curriculum. Students will be able to access the most current data by which to examine global change problems which affect the social process. By doing this they will be able to make key connections among related subject concepts and thereby enhance their ability to transfer knowledge.

Keywords. Curriculum, innovation, middle school, NASA project, Internet

1 The Nature of MINDS 2000+

The NASA MINDS 2000+ curriculum is a NASA funded project designed for middle school educators. It is built around the theme of global change and utilizes the Internet as a technological tool for teaching and learning. More specifically, educators use the integrative curriculum to address the national standards for mathematics developed by the National Council of Teachers of Mathematics and the national benchmarks for science and technology developed by the American Association for the Advancement of Science by studying issues of global change associated with NASA's Mission to Planet Earth and the U.S. Global Change Research Program.

The 'hands-on/minds-on' pre-service and in-service project is built on real life scenarios that use Internet resources to access information and to aid in the transfer of knowledge across the curriculum. Teachers integrate global change themes with the existing curriculum in order to help students develop a greater understanding of global environmental change through the concept of Earth as a system. Through these learning experiences, students explore how Earth's components and their interactions have evolved, how they function, and how they may be expected to continue to evolve.

In order to accomplish this goal, students use remotely sensed data, that is educationally relevant, captured from the Internet. Thus, the ultimate goal is to develop in students the capability to predict environmental changes, both natural and human-induced, that will occur in the future. By providing students with this capability they will be able to answer the fundamental unresolved questions about their constantly changing global environment, thus improving their quality of life.

To enhance student's ability to manage knowledge and to be prepared to teach in the future, the Eastern Michigan University's Research Institute for Space Education staff included the NASA MINDS 2000+ curriculum as part of its gopher. EMU's Research Institute for Space Education gopher (RISE) can be accessed via Internet address hardy.emich.edu. The intent of the RISE gopher is to provide a plethora of educational resources in one location. Among the items included are: key NASA educational resources and contacts; subject oriented databases; curriculum materials; and materials on teaching and learning. These items and many more can be found through the following menus located at EMU's Research Institute for Space Education:

1. About the Research Institute for Space Education.
2. Blue Skies (U of Michigan Weather Underground) <TEL>
3. NASA/
4. NASA MINDS 2000+/
5. CIESIN Global Change Information Gateway/
6. EMU NASA Space Grant/
7. Subject Resources/
8. Curriculum Center/
9. Teaching and Learning/
10. Global Change/
11. K-12 Network/
12. Other Gophers/
13. NASA Internet Resources/
14. Suggestion Box.

As a result of the project, the proposed pre-service course, NASA MINDS 2000+ has been implemented at Eastern Michigan University will in-service teacher educators and middle school teachers. A permanent institute was established for space education. Support pre-service and in-service networks were established to provide continued implementation and improvement of the quality of teaching and instruction at the middle school level. Through the established course and other products of the project, teachers, teacher educators and students will develop problem solving, creative and critical thinking skills that will provide them a means to transfer knowledge to solve tomorrow's problems, thus producing a more scientific and technologically literate citizen.

2 Learning Experiences

Based on the national standards for science, mathematics and technology the following global change themes have been selected for MINDS 2000+:

- Global Change
- Satellites and Remote Sensing
- World Hunger
- Weather
- Deforestation
- Technology Management: Urban Growth and Planning
- Technology Management: Nuclear

The curriculum map used for MINDS 2000+ is highlighted in Figure 1.

Deliverables

- Teacher's Guide (University Pre-Service and Middle School)
- Student MINDS-On Manuals (University Pre-Service and Middle School)
- Internet gopher and web server for space, global change and education

3 The Goal of MINDS 2000+

The goal of MINDS 2000+ is to develop scientific and technological literacy by focusing on science, mathematics, technology and space science as an integrator of the existing middle school curriculum. MINDS 2000+ will use this relationship and it interactions to have students develop an greater understanding of global environmental change through the concept of Earth as an entire system. Students will explore how components of earth systems and their interactions have evolved, how they function, and how they may be expected to continue to evolve. In order to accomplish this goal, students will use remotely-sensed satellite and earth probe data. Thus the ultimate goal is to develop in students the capability to predict environmental changes, both natural and human-induced, that will occur in the future. By providing students with this capability they will be able to answer the fundamental unresolved questions about their constantly changing global environment, thus improving their quality of life.

4 The Objectives of MINDS 2000+

To accomplish the intended purpose of MINDS 2000+, the project will focus on the development and implementation of pre-service and in-service activities that foster the integration of technology and space science with the existing middle school curricula for all students.

MINDS 2000+: Innovations in global change 85

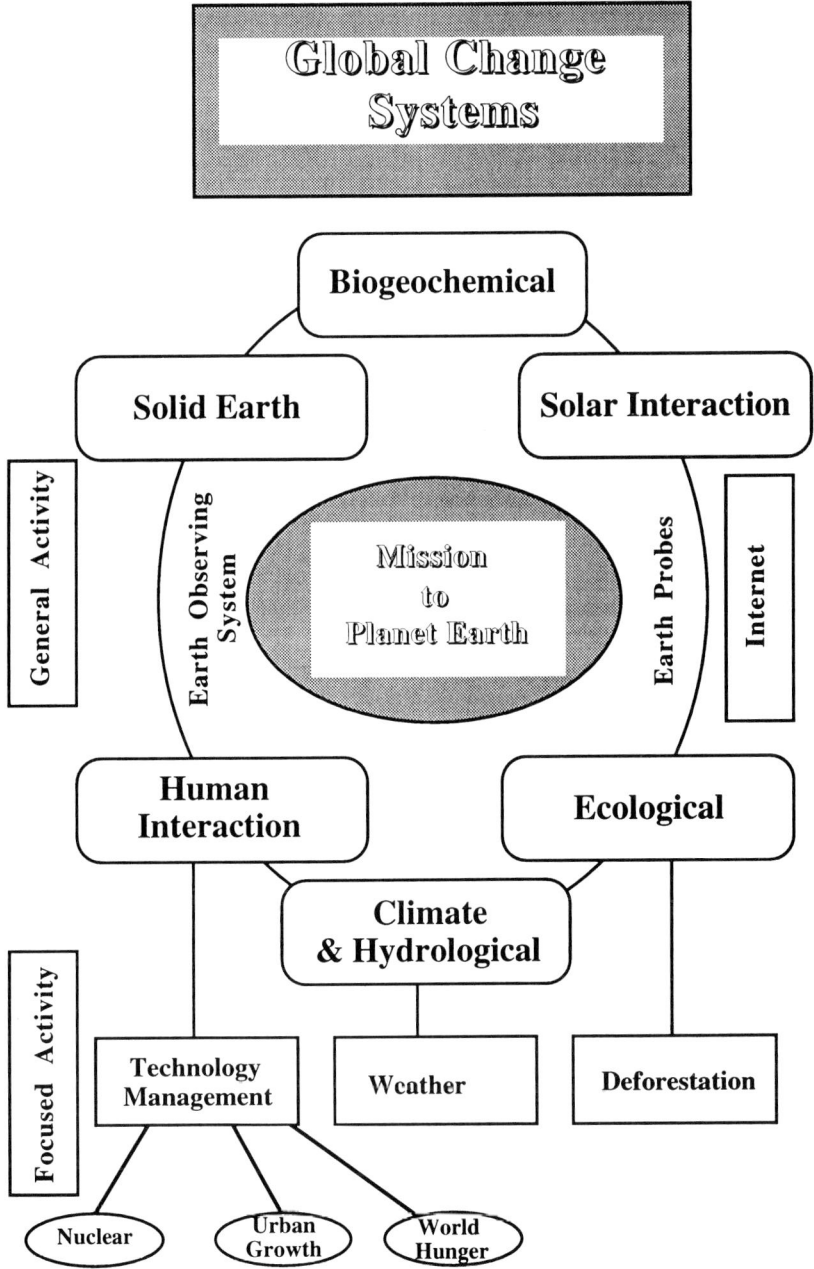

Fig. 1 Curriculum Map

More specifically, the objectives of this project are:

- to develop problem solving and creative and critical thinking skills through the integration of science, mathematics, technology and space science with the existing middle school curricula.
- to assess pupil's academic performance and teacher's instructional skills in science, mathematics, technology and space sciences as a result of their exposure to this program.
- to in-service teachers and teacher education personnel through staff development workshops and to establish a national in-service network model to enhance the transferability of the program.
- to develop an interdisciplinary model that can be easily transferred and adapted in teacher education institutions and in the nation's schools and one that enhances and facilitates the development of transferable concepts, skills, and knowledge, while enhancing attitudes and self concept.

5 The Benefits of MINDS 2000+

MINDS 2000+ fosters the following benefits to students:

- National Standards for Science, Technology and Mathematics
- Interdisciplinary Approach
- Systemic Approach
- Innovative Teaching and Learning Practice
- Use of Internet and other databases
- Problem Solving and Problem Finding
- Creative Thinking
- Critical Thinking
- Collaborative Learning
- Interpersonal Skills
- Real World Context
- Environmental and Social Awareness

6 Internet and Learning Experiences

Students work in groups to solve global change design briefs. A sample design brief in illustrated in Figure 2.

MINDS 2000+: Innovations in global change 87

Deforestation

People are creative because they make a conscious effort to think and act differently. All people have the ability to be creative.

The problem outlined below provides the bare minimum of information you need to proceed with the design process. While you must think about the actual implementation of your design, you are only asked to complete a model which represents your solution. It is important that you plan a strategy, and consider many alternative designs before working on the one you think most appropriate. Let your minds go free, and try not to be too bound by rigid thinking.

PROBLEM SCENARIO: Over the last several decades increased pressure has been placed on the world's forest resources. The cry has been for more effective forestry management. One of the areas most affected by this demand is the Pacific Northwest. The Pacific Northwest is in the midst of a battle between the timber industry and the environmentalist. Questions are being raised by the timber industry for rights to clear cut forest areas, especially public forest, in order to protect the industry's jobs and to lower the cost of building materials due to a greater supply of lumber. Among the leaders in this effort by the timber industry are the Northwest Forestry Association, the lumbering communities, and the companies. Environmentalist are trying to block the efforts of the timber industry in order to protect the decreasing amount of old growth forest, the global carbon cycle and endangered species, the spotted owl. Among the environmentalist groups are the U.S. Forest Service, the Bureau of Land Management, the Wilderness Society, the Oregon Natural Resources Council, the Audubon Society and the American Forest Service Employees for Environmental Ethics (AFSEEE). Who is right? Can a forestry management plan be developed and implemented that will bring a balance between the spotted owl, old growth forests and the timber industry's need to harvest trees on public land?

**Genius is one percent inspiration, and ninety-nine percent perspiration.
Thomas Edison**

DESIGN BRIEF: The class will be divided into two groups: timber industry and environmentalists. Each group will create a forestry management plan in the form of a report that will allow the timber industry to be productive and at the same time be environmentally sound. In order to accomplish your task, each group will be subdivided into the following groups: natural scientists in charge of observable elements, physical scientists and technologists in charge of atmospheric transmission, sensors and satellites, information technologists in charge of data and information processing, and local, state and national social scientists in charge of creating policy. All work will be recorded in the group's portfolio. Group members will record all their work in their design log. Each group will construct a model to depict their solution. Once each group has completed their research and produced their model and report, the groups will debate their findings and viewpoint. This debate will be judged by local community experts.

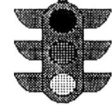 *You may now proceed with wild abandon...............*

Fig. 2 Deforestation design brief

Groups are organized based on student interest and learning style. Learning experiences involve the interactions among natural science, physical or biological sciences, information technology and social sciences. This interaction is diagramed in Figure 3.

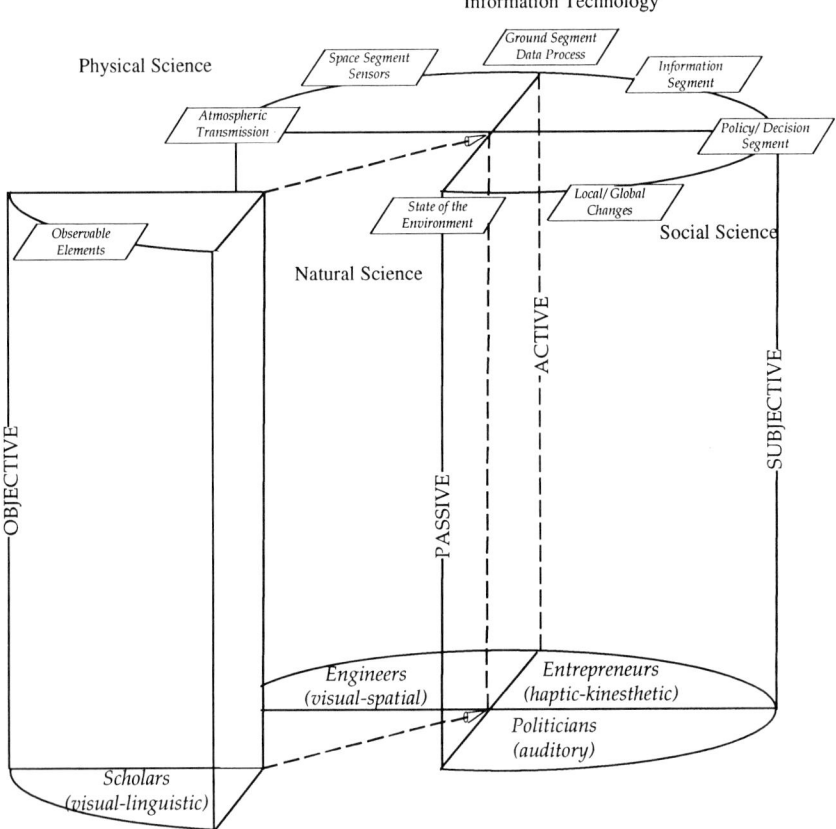

Fig. 3 Systematic global change model

In order to solve the global change design briefs, students use Internet as an integral component of their learning process. They use primarily the Eastern Michigan University NASA gopher and web, the Consortium for International Earth Science Information Network's (CIESIN) gopher, the University of Michigan's Underground Weather database (Blue Skies), NASA Spacelink and Internet resources. Through these resources, students are able to capture key governments documents, articles, demographic information, and images. In order to gain a better understanding of the type of research categories available to students in the menus of the Eastern Michigan University NASA gopher and web, which is highlighted in Table 1.

Table 1 Eastern Michigan University's RISE Gopher

1. About the Research Institute for Space Education (RISE).
2. Blue Skies (U of Michigan Weather Underground) <TEL>
3. NASA/
4. NASA MINDS 2000+/
5. CIESIN Global Change Information Gateway/
6. EMU NASA Space Grant/
7. Subject Resources/
8. Curriculum Center/
9. Teaching and Learning/
10. Global Change/
11. K-12 Network/
12. Other Gophers/
13. NASA Internet Resources/
14. Suggestion Box.

Students begin MINDS 2000+ by study general concepts of global change. As part of this study they go on to a scavenger hunt of the Eastern Michigan University NASA gopher and web. The scavenger hunt enables them to learn about Internet and to become familiar with what data is available to them on global change. They also explore ftp's (file transfer protocols), E-mail, telnets and other Internet resources. Once students complete this learning experience they explore satellites and remote sensing. They study the Earth Observing System and Earth Probe satellites. In this learning experience students use Internet to collect data on each satellite as to its operation, platform, resolution, swath and purposes.

With this background students embark on a study of world food problems. Students are divided into the UNESCO geographic regions in order to study a particular region's food problems. Besides the CIESIN gopher, students use the geographic server (martini.eecs.umich.edu 3000) to gain information on cities, latitude/longitude, food resources, imports, exports and the like. They also access map data through ftp spectrum.xerox.com and the CIA World Data Bank through ftp gatekeeper.dec.com and ftp ucsd.edu. An example of a image capture is found in Figure 4.

Students then go through a series of weather-related activities. In this learning experience students learn basic weather concepts, capture weather images, and study disaster management of weather catastrophes. There are at least 40 ftp's and telnets for weather information and images. Among the most useful are Blue Skies (University of Michigan Underground Weather), NOAA, NCDC Weather Data, ftp uriacc.uri.edu for weather images, ftp unidata.uarc.edu for weather images and aurelie.soest.edu.au for sea-surface-temperature.

In the learning experience on deforestation, students are divided into two groups: (1) loggers and (2) environmentalists. Students will study deforestation from the viewpoint of their group. They will access CIESIN's gopher, use the Carbon Dioxide Information Analysis database and other Internet databases.

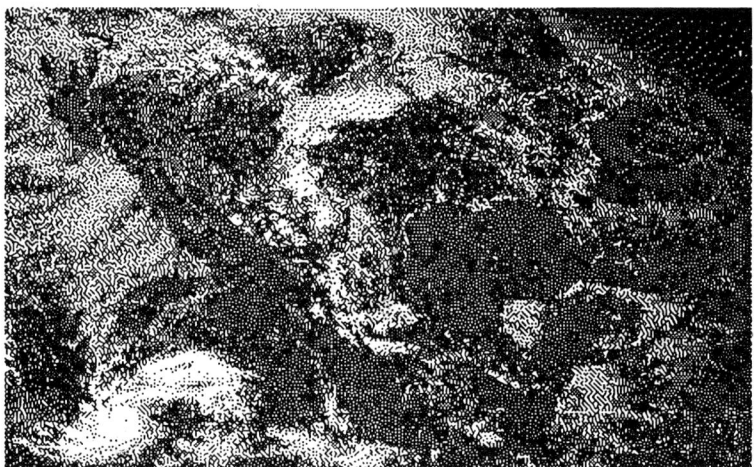

Fig. 4 Image capture example

Students will use the data accessed to design a case for their viewpoint. They will end the learning experience by E-mailing their class discussion to their representative and the White House.

Students will study the global change problems associated with the urbanization of large international cities through the problem on Technology Management: Urban Growth and Planning. They will use the geographic server (martini.eecs.umich.edu 3000) to gain information on cities, latitude/longitude, food resources, imports exports and the like. They also access map data through ftp spectrum.xerox.com and the CIA World Data Bank through ftp gatekeeper.dec.com and ftp ucsd.edu. Besides these Internet resources, they will access reports and articles from libraries, such as the Library of Congress.

Finally, students will study Technology Management: Nuclear. This learning experience explores the energy component of global change. Students will access the CIESEN gopher, the National Nuclear Center database, the CIA Data Bank and other necessary telnets and ftps. They will capture images over time, i.e., 1986, 1988, and 1992 to examine the changes that have occurred around Chernobyl.

Key Qualifications - Contents of General and Vocational Education

Manfred Lutherdt
College of Education Erfurt
Institute of Technical Sciences and Company's Development
Nordhäuser Str. 63, D-99089 Erfurt, Germany

Abstract. Key qualifications in commercial and administrating fields are thinking in hierarchical systems, interpersonal communication, interpersonal interaction and self-management. Key qualifications in technical fields are ability of problem solving/creativity, ability of thinking (system thinking, hierarchical thinking) and ability of valuation and judgement.

Keywords. Key qualification, motivation, re-employment, vocational education

1 Globally, the Changes Regarding Social Values, Regarding Economy, Science and Technology Are Discussed

In connection with the demographic changes, the level of production, the wage costs, the free competitions on the foreign markets, the innovation ability of the production, etc. the consequences are different in each country. But: High qualification and productivity of people – as an inevitable precondition – can and has to be secured everywhere.

i. The innovations themselves and the consequences of the development call for rethinking in education. A lifelong-lasting education - once acquired at school and at the vocational training - will be only a dream. In future, the ability for a lifelong self-education will be the dominant feature of personality. Selfmanagement will be a task because up to now it was only taken seriously for leaders.

ii. The changes of working fields, working contents and work organization produce another situation.

– Technical means, technological processes determine increasingly the picture of the traditionell commercial, administrative and service oriented sectors.

Technological knowledge, abilities and skills and the creative dealing in using the technique is therefore a necessary precondition in all professional spheres and spheres of life.
- The rapid changes of the working contents and the work organization force to the reprofessionalizing and reconcretizing of the professional qualification processes.

The phenomenon of fast devaluation of professional knowledge is not new.

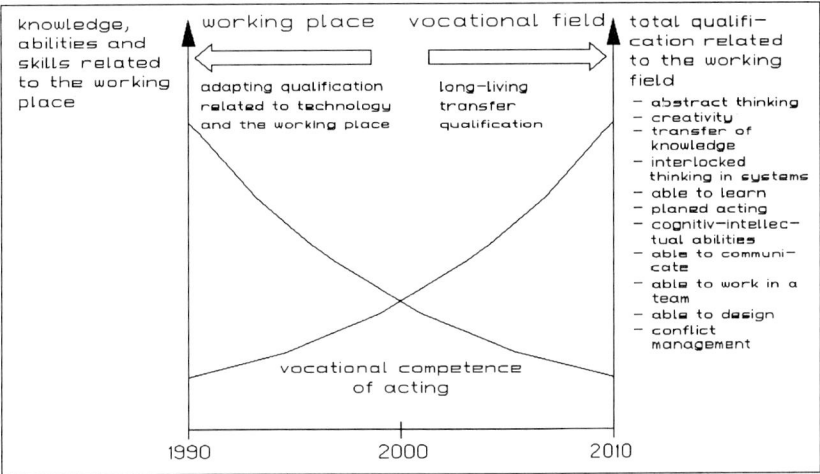

Fig. 1 The development of working place related knowledge, abilities and skills and working field related total qualification

The interlocking of the single and general by the special was solved in the general education by the approach of invariants.

Regarding the technological education the determinant invariants are no longer only derived from the technical objects and systems but also from the technical stategies, methods and processes.

- Thinking in evolutional systems (interlocking, hierarchy)
- Problem solving ability and creativity
- Searching for alternatives
- Compromise ability as a result of valuation ability and explanation ability and others are educational aims derived from these invariants.
- The paradigms change from the single person to the communication andinteraction in teams is another result of the worldwide development.

Taking into consideration the developed-technical systems interpersonal communication and interaction are further moments, which have to be considered for the determination of the educational aims and the didactics.

2 Which Consequences Does This Development Have for Education?

i. In the partial systems:

- primary school / preparatory school
- secondary school / high school
- vocational training and
- occupational system

the reactions on the developments differ in time and in content. It follows that the already existing contradictions between the systems increase. Obviously, the shaping of the ecological transitions is not enough above all from the general education to the vocational education and to the occupational system. For a long time the possibility existed to select equal representation of educational and work objects at school and at the occupational system. The innovation itself dropped this approach.

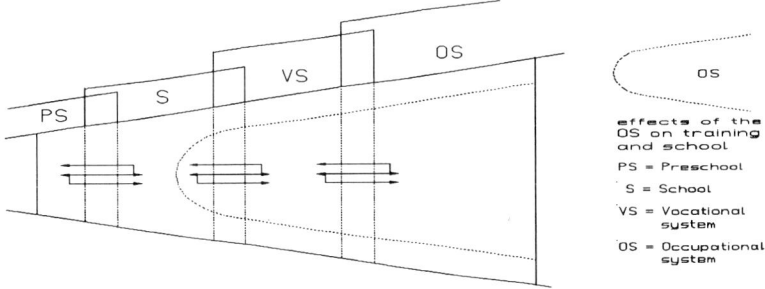

Fig. 2 Cone of development: school – occupational system ecological transitions by interaction and optimized decentralized feedback

ii. The result of the relative isolation of the general education towards the occupational system with its differentiated companies and industrial sectors is felt by the people themselves. Unemployment, particularly by women, is one of the results of the insufficient compromises between general education and the vocational demands of the occupational system. Rationalization is experienced as a process of exchange of production factors.

Technological systems, automatic production processes and computerized process controls and information realization are not only improving technological possibilities of production. They are also the rational possibilities of production in the combination process of the factors at this level of production, if they are combined with the adequate productivity of manpower.

By planning the production taking into consideration these premises you will have lower costs on the market in comparison with the competitors. This creates security, chances of survival of the enterprises and wealth of the nation.

3 Is There Any Approach of Solution?

The sociology of industry has discussed some approaches. The soundest approach seems to be 'the approach of ending the division of labor with the imparted key qualifications'.

The theory of 'overqualification' and the practice of the at present existing dual training lead to the fact that manpower is on one hand too expensive and on the other hand people are not motivated to carry out simple work. The theory of 'dequalification', a variant of the exploitation thesis, proved to be one-sided similar as the thesis of 'overqualification'. Both theories were refuted empirically. With the theoretical approach to the 'key qualification' the possibility will be opened to exert qualifications in modular structure determined by the technological change. The global aims exist in enabling somebody to the 'paradigms change'.

Table 1

from	to
from the mechanical – linear	to the integrated interlocked thinking
from the experienced – conscious know – how	to the situational analysis and to the creative problem solving
from the activity – specific	to the functional – spreaded mastering
form the valuation of the single information	to the complex valuation and explanation
from the immediate solving	to the future – oriented solving of problems
from the single persons	to the communication and interaction in teams (groups)
from the outside management	to the self-management regarding learning and working

4 What Are the Contents?

Key qualifications are not the expression of outer-professional demands. They are rather based on an intensified understanding of technological processes and objects, technological way of thinking and acting - thus referred to the above an understanding of up-to-date skilled work.

Thus, the vocational basic knowledge, skills and abilities are the basis, the medium for determining key qualifications. The following key qualifications are mentioned for the commercial-technical sphere.

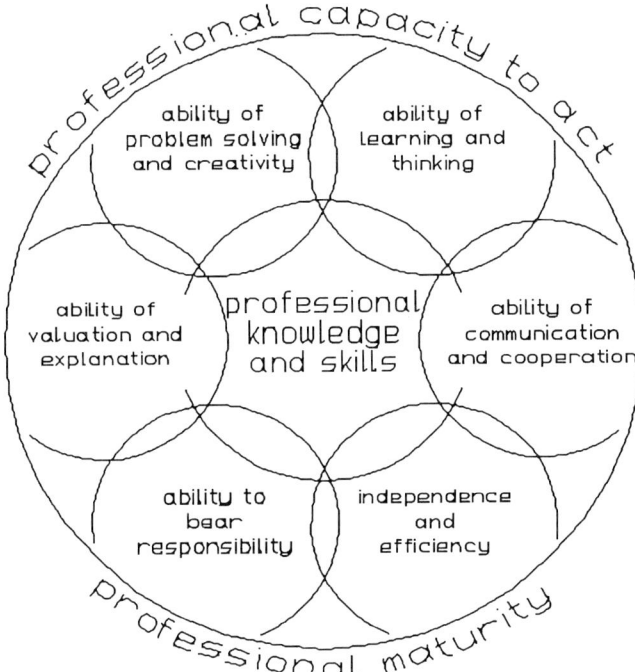

Fig. 3

By that, traditional technological specialized qualifications are not depreciated. They will become the basis, the representative, the source of key qualifications.

Table 2 Different learning content of different task

task up to now	reading the drawing	problem analysis	alternative task
'producing a workpiece according to the drawing' given: drawing, parts list, single work		to define the criteria of judgement/demands	'planning and producing a piece which shall meet special functions and demands' given: the problem, measurements within a given scope (limits), according to a drawing, team work
		to determine/to work out general solutions	
		to work out alternatives/variants	
		to find out experimentally detail solutions	
		to inform about materials, connecting elements, working processes	
		to valuate and to select alternatives	
		to determine measurements and details	
		sketching and drawing	
	to plan the working steps	to plan the working steps	
	to inform about tools, materials, machines,..	to inform about tools, materials, machines,...	
	proper use of tools, machines,...	proper use of tools, machines,...	
	measuring and checking	measuring and checking	
		optimizing of partial solutions	

result: a workpiece according to the drawing with different qualities		to valuate the results according to the demands	result: different workpieces meeting the demanded functions in different variants
	visual and measuring control	visual and measuring control	
		to discuss solutions, articulating and defending of opinions	

Within the commercial-administrative sphere we have tested successfully the following key qualifications in a two-year research assignment.

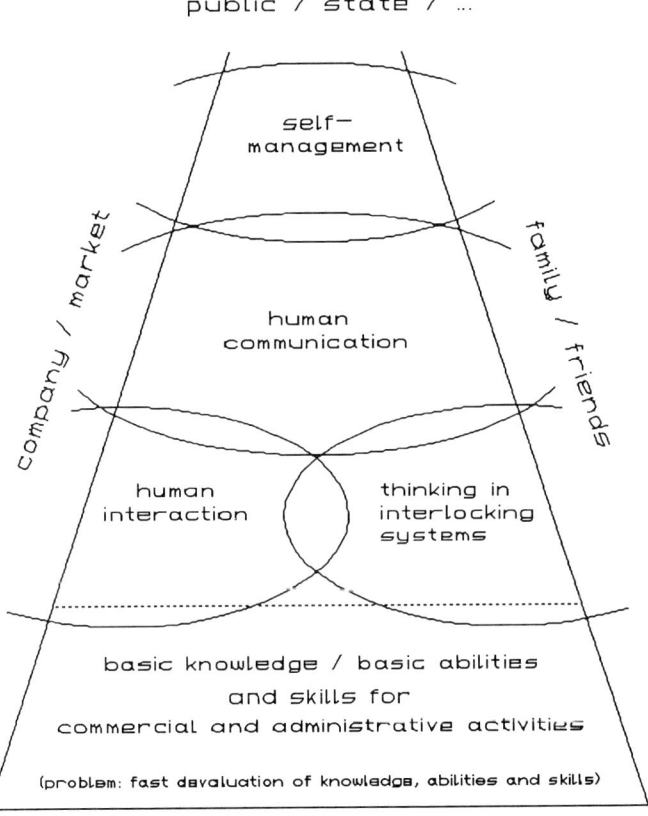

Fig. 4 Pyramid of qualification

The target group were women with an avarage age of 50 years. After finishing the one year qualification, nearly 100% of these women were re-employed, in spite of 15 to 18% unemployment in the region, of which 61% were women.

Starting point of our considerations had been the radical changes after the reunification, also relating to the occupational system.

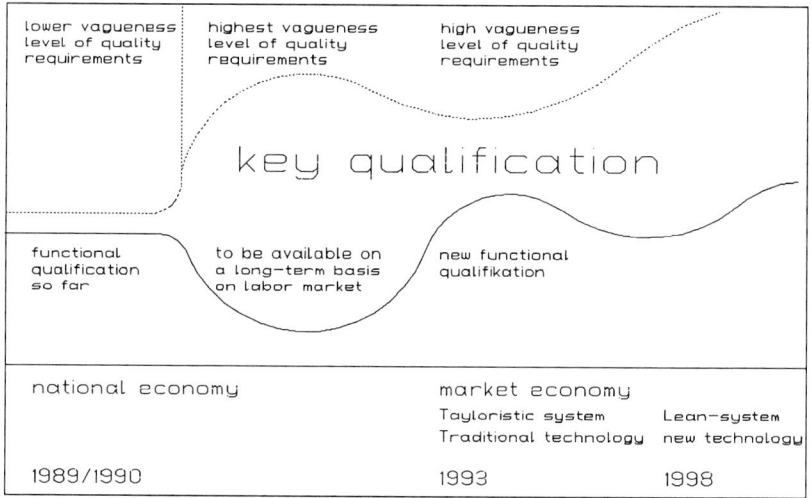

Fig. 5

We have seen in the new countries of Germany and in the countries of Eastern Europe a change of values (standards) for which somewhere else up to now one or more generations have had time. The phase structure of qualification is represented on the next figure.

The modulare contents M1, M2, ... , M7 are realized by a didactics oriented at the target group and at the aim of qualification.

Entrance phase
- motivation and orientation
- the female participants - in spite of their differentiated knowledge, abilities and skills - will be led to a uniform level.

Main phase
- internal interlocking of the content of the modules; preparing the first practical training
- first practical training
- interlocking the contents of the modules
- second practical training
- project work as an accompanied demand

Result: By practical learning, systematic thinking and self-controlled acting a high exit level will be reached.

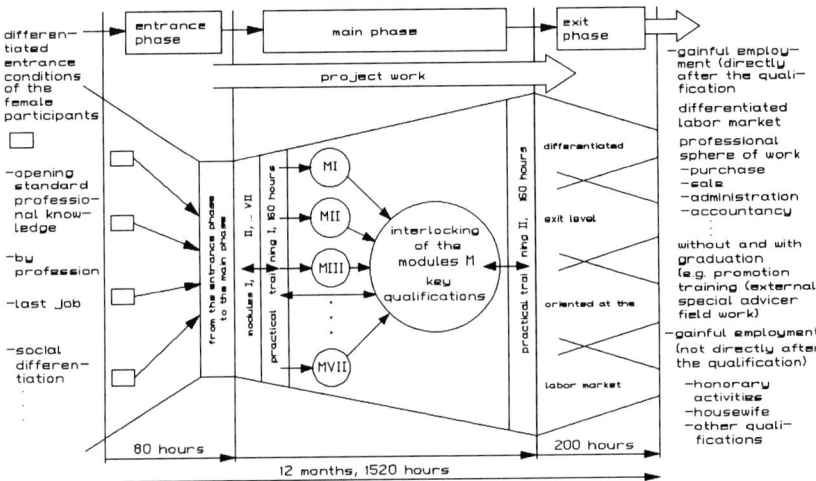

Fig. 6 Phase structure of qualification

Exit phase
- individual adaptation to the working activity in the different fields of the area (sector)
- special functional knowledge, abilities and skills

The interaction oriented approach formed the foundation of a didactics for the appropriation of key qualifications. Such didactics have been developed.

5 Key Qualifications - a New Approach to Determine General Technological Training

The concept of key qualifications characterizes a movement, which breaks away from the high-differentiated and intricately handled approach of qualification to a general catalogue of demands, which approaches to structure of general aims of training.

The approach did not only come out since the technological general education took over more and more vocational contents but because the vocational training has to take up the consequences of innovation. This development will decrease the differences between general and vocational training. Technological training is more than key qualification.

But: The invariants will remain representatives as technological systems and processes of the world of living and working and the invariants resulting from the peculiarity of the purpose-oriented technique shaped by man. The last mentioned are nearly equal with the key qualifications. Therefore, it seems to be the right way?

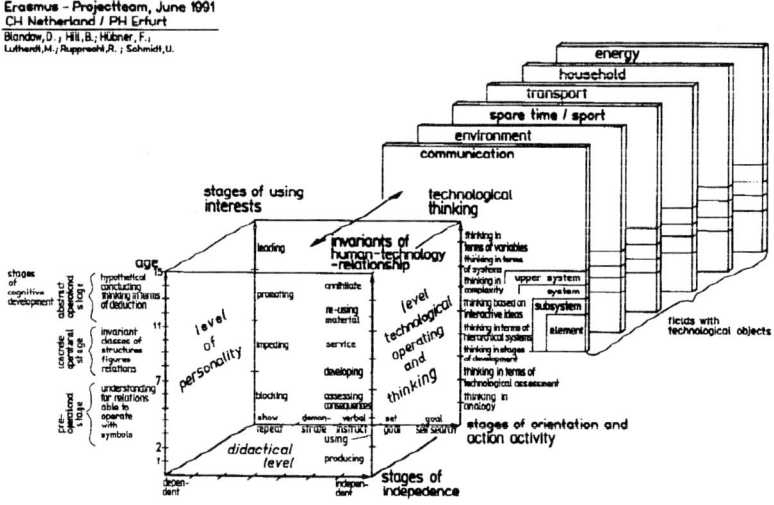

Fig. 7 Morphological box for determination and selection of pilot projects

References

1. Lutherdt, M.: Qualifizierung und Neuorientierung von Frauen für kaufmännische und verwaltende Tätigkeiten. Pädagogische Hochschule, Institut für Technische Wissenschaften und betriebliche Entwicklung, Abschlußbericht, 1994
2. Beinke, L.: Qualifikation und Neuorientierung für Frauen im kaufmännischen und verwaltenden Bereich in Thüringen. Gutachten. Gießen, Verlag Arbeitslehre – Didaktik, 1994
3. Theuerkauf, W., Weiner, A.: Key qualifications as an ability for a Technical Education. In: Technological Literacy, Competence and Innovation in Human resource Development. Proceeding of the First International Conference on Technology Education. Weimar 1992
4. Sachs, B.: Schlüsselqualifikationen in der Berufsbildung und im allgemeinbildenden Technikunterricht. tu 3/1993 pp. 5-12, (1993)
5. Lutherdt, M.: Basic patterns of technological education in Germany, the theory approach ... and the innovative theoretical approach to an independent technology instruction. JIEA – PATT – International Conference, Reston, Virginia 10/1992 S. 78-98, (1992)

Strategies, Methods and Principles of Development Thinking

Bernd Hill
College of Education Erfurt
Institute of Technical Sciences and Company's Development
Nordhäuser Str. 63, D-99089 Erfurt, Germany

Abstract. Technology education is pointed towards the ability of designing according to the variability of technological present and the conception of the future. A condition for the designing of technology is the shaping of development thinking. Development thinking is one from the level of technology as well as from the anticipation of a vision a feed-back thinking for the development of:

– more effective
– more pollution-free and
– more human

technical artifacts. Development thinking does not only presuppose the knowledge about strategies and methods but also the knowledge about heuristically used principles which lead to the more concrete technical solution. Therefore a development understanding in technology education can be marked in connection with the ways of action and laws of evolution for the determination of goals and the principles of problem solving.

Keywords. Development thinking, technology education, bionics, problem solving methods

1 Introduction

A country that is poor in raw materials like Germany needs a creative potential, which determines the quality of the technical future by ability of innovation and self-responsibility.

As the technical world is and has to look like, belongs to the shapable possibilities of a human being. Therefore ability of innovation based on creativity has a key-competency in this process of designing.

Technology education has to include the dynamic character of technology because technology itself has to be understood as a developing process and not only as totality of closed temporal structures. The technology lesson has to offer possibilities to enable pupils for development of technology in society. The aim will be to change and to develop technology more pollution-free and more human. If one looks at the variety of relations between a human being and technology the following educational intentions support their realization:

- technology because of its multidimensionality presents a compromise of contradictive aspects (Blandow, 1992).
- thus technology means human doings with an aim of combining nature-like possibilities with economic/ecologic logic and human/social wants (Traebert, 1990).
- technology is a goal-oriented effective changing doing and therefore a spatial-temporal change of structure, realized by technological behaviour.

Therefore technology in a lesson:

- must be made obvious as a complex entity with manyfold interlinks seen from an ecological, economical and social point of view by aspects of complexity and complication
- must be consciously formed by more human and nature-oriented principles
- should be comprehended in its various forms of appearance as a process of development

In the foreground stands the independent and original mastering of technically directed life situations by an elementary technical problem analysis to gain a competence of designing directed to combine technology with nature.

Technology lessons have to include the appropriation of basic pattern of development, to gain insight for transferable correlations and principles as a precondition for the comprehension and changeability of the technical world. Under these conditions technological education includes the strengthening of development thinking.

2 Features of Development Thinking

Technology itself represents a process of development, a historic genesis upto the present days and includes the anticipation of the future (technical changes).

Each technical system seen from a technological point of view embodies a moment-constellation. Technology itself represents a spatial-temporal process of development as a branching development line of systems. For this reason the resulting development thinking is also spatial-temporal implanted. It comprises above the present stage of technology the historical genesis of technical phenomena upto anticipation. The anticipation of functionable technical structures is not merely spiritual anticipation of technology but it also includes visions (how technic could look tomorrow).

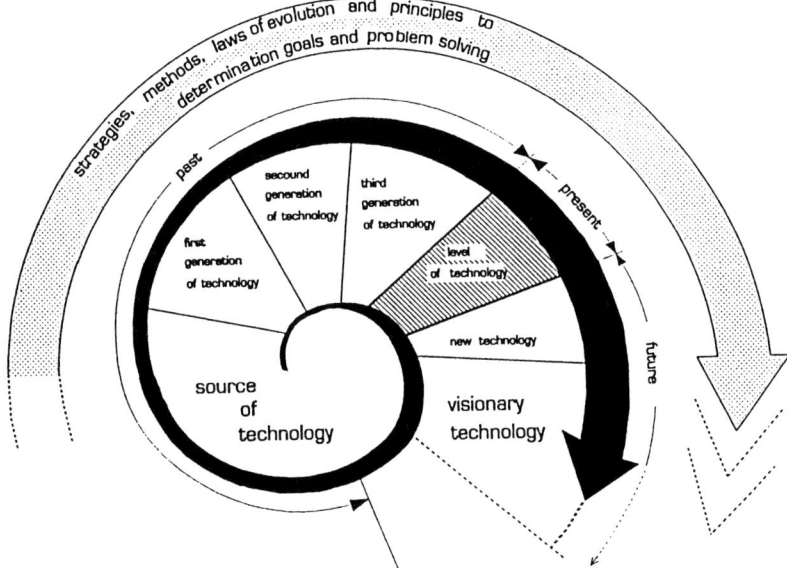

Fig. 1 Spiral of evolution

Development thinking seen from the technological point of view as well as considering future visions of anticipation is a feed-back kind of thinking for the development of:

- more effective
- more pollution-free and
- more human

technical artifacts. Development thinking includes furthermore the system overlapping aspects of the usage of biological systems as for design ideas and the starting point for technical solutions. The ability to determine and formulate goals because of their multidimensionality and to realize them as a novel problem solution in technological results needs the strengthening of development thinking.

3 Foundational Orientation for Effective Strategy Forming

To strengthen the development thinking it needs proper teaching and the appropriation of directional aids for the development of technology. Such directional aids as strategies and methods by goal-oriented and systematic use facilitate an effective problem recognition and problem solution. During the process of solution finding one has to overcome thinking barriers to reach from the initial state (problem situation) to the final state (problem solution). Thus a

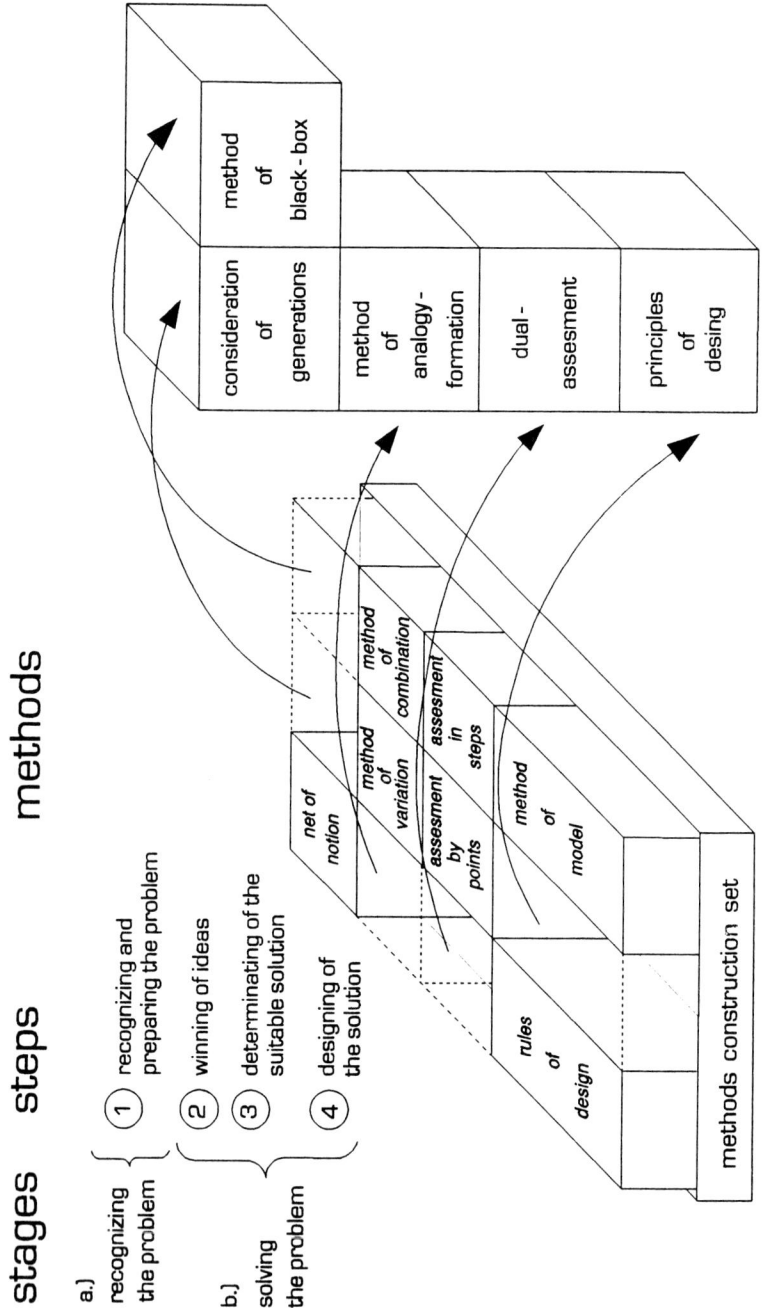

Fig. 2 Methods construction set

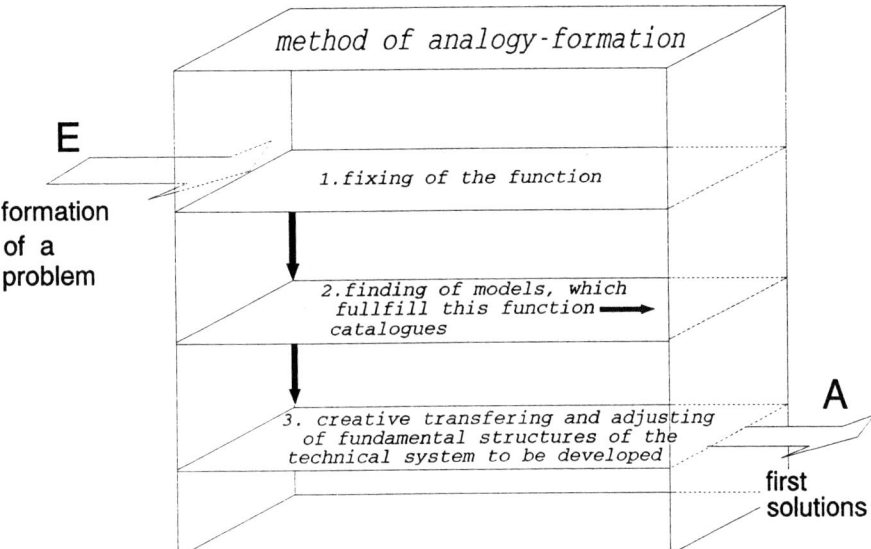

Fig. 3 Structural stone: method of analogy-formation

conscious usage of methods can help to overcome thinking barriers more effective and provide space for creativity.

In a technology lesson the intentional use of methods of problem recognition and problem solution as appropriation items require their preparation from a didactical point of view. For these methods relevant usage features should be fixed and pointedly marked; the methods to be allocated to the steps of the process of problem solution, consequently the relevant methods for each step to be summarized as a method group. These methods can be provided in ordered criteria to the pupils as a 'Method building kit' (Hill, 1994).

The ability for an independent method competence will exist if the problem solver can select the methods according to the appropriate problem structure and use the 'bricks' of the 'method building kit' for the creation of an up-ranked action-programme.

As a result of this selection process a linear branching method order is created as operational structure of technically developed abilities.

4 Analogy Classes for Goal Fixation and Solution Finding

Inspiration plays an important role for generating a solution approach. Intuition can be controlled by appropriate methodical means. Such methodical means have to meet the demand of a goal-aimed release of associations.

Analogy classes consist of heuristic utilizable catalogues which contain beside representatives in model form usable principles and exposition of structure. The usage of these catalogues as a source of inspiration can facilitate respectively accelerate the problemsolver's intuition. Therefore for the goal determination it is possible to derive development potentials from evolution regularities. This is underlined by the transfer of these regularities upon the current state of technology.

For the forming of solution ideas analog solution principles of nature and technology are usable. By generating solution approaches the catalogues belonging to the analogy class for can be examined by its content of ideas to solve the problems. For example a derivasion of effects and principles from biological systems might be heuristically useful for a solution finding. A multitude of catalogue leaves are allocated to the single principles, containing the structure exposition of the biological systems. The thereby won principle structures demonstrate basic solutions, which by reason for realization of demands and conditions by variation and/or combination of available features and a final evaluation, lead to a solid technological solution.

The usage of evolution regularities for solution determination and of principles for the solution finding opens new directions of development and efficient solution possibilities for the problem solver's design work of technical artifacts.
In addition they are new elements of technological lessons which are directed to development and design.

References

1. Blandow, D.: The Elements of Technology Education. TU Eindhoven, 1992
2. Hill, B.: Bionics – Element of Inspiration in the Process of Problem Solving. In: tu Technik im Unterricht. Villingen-Schwenningen, Neckar, 1994
3. Traebert, W. E.: Technology and General Education Schools. In: Technik und Philosophie. Düsseldorf, VDI, 1990

Cognitive Structure of Technology Subjects for Deciding Instructional Design and Learning Preferences

S. Swaminatha Pillai
Technical Teachers' Training Institute
Department of Educational Research
Taramani, 600113 Madras, Tamilnadu, India

Abstract. Cognitive mapping leads to knowledge structures as applied to subject disciplines taught in educational programmes and institutions. Cognitive, Conative and Affective domains are considered as dominant characteristics (or dimensions) of the subjects of study, methods of instruction and personnel in education. An attempt to consider at least three levels on each dimension is made to classify the above three educational components in a 3-dimensional structure. Implications for curriculum development and teacher education are brought out.

Keywords. Knowledge structure, cognitive mapping, psychological dimensions (cognitive, conative, affective), instructional design, learning preferences

1 Cognitive Science and Technology Education

Cognitive science as a modern influential form of psychological approach to human knowledge has established itself strongly in the field of education. The two cardinal principles of individual responsibility and constructed meaning (O'Neill Jr. and Spielberger, 1979) have wide implications for educational development. Two of them indicate the need for further research on attribution of success or failure in learning to a changing internal cause such as effort and the reality of differing mental processes in learners constructing mental elaborations of the matter to be learnt. Such efforts and elaborations of students in technology education are in consonance with the nature of the scholastic discipline called technology. The foundational knowledge and basic skills acquired through science education are made more productive in technology education as denoted by the distinction between Science and Technology. Inasmuch as the productivity for concrete outcome of technology education is rooted in intellectual readiness of students, cognitive science contributes to technology education.

1.1 Cognitive Mapping and Knowledge Structures

The concern of a cognitive scientist is with how an individual "...selects, codes and uses information" (Wood, 1983). Human cognitive processes are dependant on mental representations postulated by schema models of knowledge. Ausubel's Cognitive mapping, Piaget's schemata and Kelly's Personal Constructs point to knowledge structures in the form of 'encyclopedias rather than dictionaries'. An object or an incident is usually described by an individual in the form of his pattern of organising incoming information, in other words by using a script or schema particular to him. "Schemata are generic knowledge structures that guide the comprehender's interpretations, inferences, expectations and attention" (Graesser and Nakamura, 1982). For example, a road accident witnessed by a traffic inspector, an automobile engineer, a highway engineer, a lawyer and a poet would report describing it using the individual schema. In Piaget's view: "... in any cognitive encounter with the environment, assimilation and accommodation are of equal importance...". His model of human cognitive system stresses the constant interaction or collaboration of the internal-cognitive with the external-environmental in the construction and deployment of knowledge (Flavell, 1977).

Martin (1984) proposes three levels of embedded cognitive schemata conceived as exerting executive, metacognitive and cognitive control in the selection of "knowledge structures/networks that, in turn, instantiate the schemata, thus adapting them to ever changing situational information".

1.2 Information Technology and Education

Scripts and schemata are then larger units of information processing necessary to deal with context effects and understanding (Wood, 1983). "Information technology deals with communication, storage, processing and use of information for a variety of valuable uses." (Sadanandan and Chandrasekar, 1987). If the five senses are gateways to knowledge and the sixth sense of rationalism is typically human, there is the seventh sense of metacognition useful in human learning. Nisbet and Shucksmith (1988) focus on 'learning to learn', a topic of central interest in the area of Cognitive psychology. "Metacognition refers to one's knowledge concerning one's own cognitive processes and products or anything related to them."

1.3 Educational (Information) Technology

Dede (1985) implementing the new information technologies in education, identifies the following effects of New Educational (Information) Technologies (Forester, 1985).

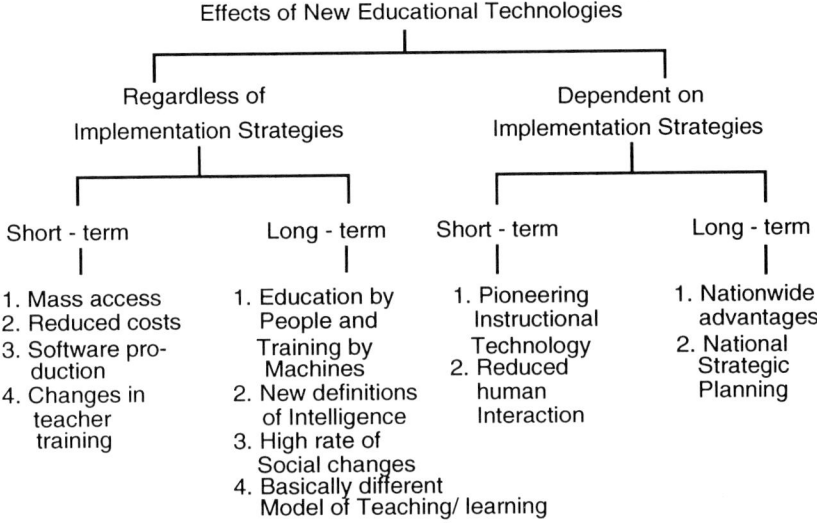

Fig. 1 Effects of new educational technologies

Thus the intertwining of information and educational technologies proves to be a provider of educational advantages both long-term and short-term, regardless of as well as dependent on implementation strategy. The constantly widening scope, meaning, significance and value of educational technology to establish itself as a full-fledged branch of technology, provide for it the pride of place in technology education, management and innovation.

Jamal and Se-wun (1993) identify factual knowledge, cognitive complexity and frame repertoire as related cognitive units used in processing complex information. According to them, fragmented and disintegrated data gathered from mass media and other sources and the classified factual knowledge are stored in the 'Cognitive bins' called conceptual frames strongly related in an individual's repertoire attributed as cognitive complexity. Scholastic subjects are also cognitively complex.

2 Knowledge Structure of Scholastic Subjects

Subjects of studies (disciplines) as storehouses of knowledge can be classified according to their frames of factual knowledge. The current considerations can be presented as Fine Arts, Performing Arts, Humanities, Social Sciences, Basic Sciences, Applied Sciences and Technologies. The frames of factual knowledge in each group of subjects can have more or less of the three domains of learning, viz., Cognitive, Conative and Affective. Cognitive complexity is characterized by various combinations of these domains in each subject of study. For example, fine arts per se such as painting and sculpture are highly affective, less conative and least cognitive. Performing arts are highly conative, less affective and least cognitive. Philosophy and Literature as humanities are equally cognitive and affective and less conative. Social sciences as diversive as psychology and economics are equally cognitive and conative but least affective. Basic Sciences such as Physics and Biology are highly cognitive, less conative and least affective.

Applied Sciences such as Geology and Metallurgy are equally cognitive and conative but less affective. Technologies like Engineering and Textile are highly conative, less cognitive and affective. Such descriptions can be further narrowed by adopting semantic differential technique.

2.1 Dimensions and Levels

An attempt is made in the following graphic representation of the three domains as dimensions and at least three levels of complexity. Keeping cognitive domain on X-axis, Conative domain on Y-axis and affective on Z-axis, the placement of scholastic subjects in such a figurative cube may appear as in Fig. 2.

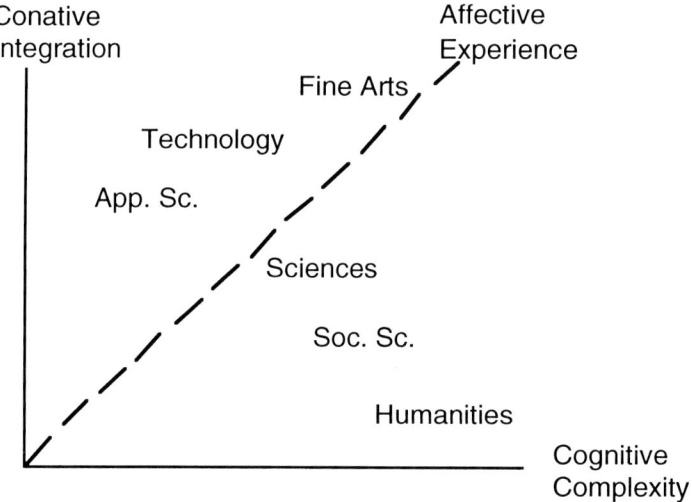

Fig. 2 Three-dimensional Description of Scholastic Subjects

Although each dimension can have different levels it is proposed that for the sake of uniformity each dimension can have three levels of simple, moderate and complex (for cognitive dimension), easy, average and difficult (for conative dimension) and direct, covert and angular (for affective dimension). The knowledge structures of all scholastic subjects are thus accommodated in a 3x3x3 cube for 27 major disciplines to be further analysed into more and more subjects as human knowledge continues to expand.

Elaboration of the graph would yield the following cube:

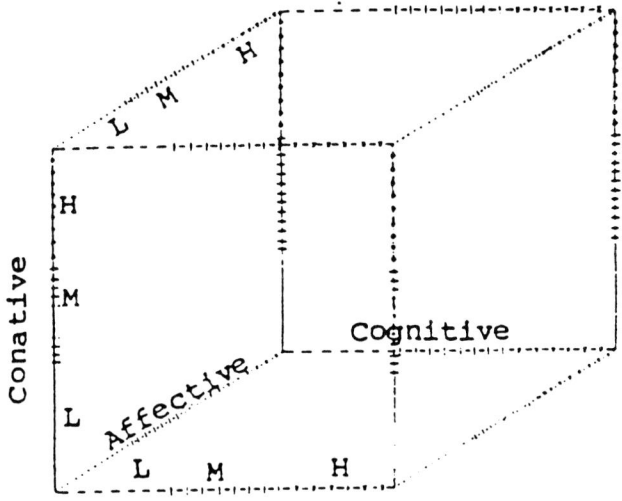

L = Low   M = Medium   H = High

Fig. 3 Dimension & Levels of Scholastic Subjects

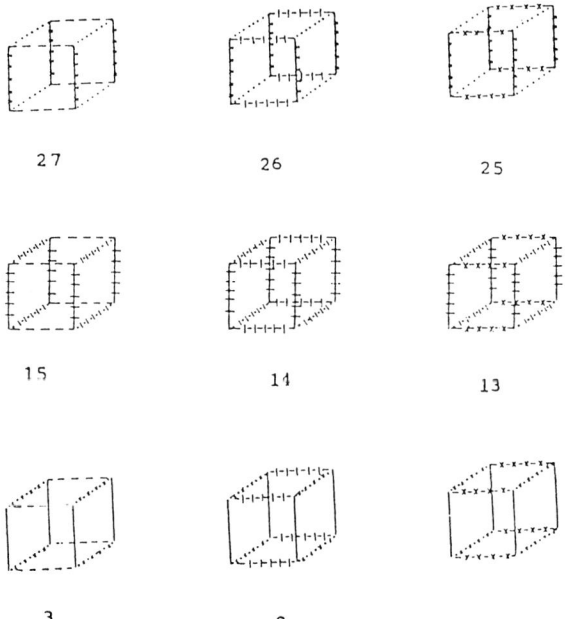

Fig. 4 Cubes Extricated from the Structure to Illustrate Specific Subjects

The twenty seven cubes shown above represent 27 combinations of the domains and levels to indicate the nature of small groups of scholastic subjects or mostly one major subject discipline.

3 Instructional Methods and Their Characteristics

Winn (1990) offers three implications of cognitive theory for instructional design: "...Instructional Strategies need to be developed to counter the reductionism implicit in task analysis; design needs to be integrated into the implementation of instruction; designers should work from a thorough knowledge of theory not just from design procedures". Pirolli (1989) put a new instructional design into practice on the art of building. Pirolli and Russell (1990) presented the Instructional Design Environment, a hypermedia system for designing and developing instructional material, including texts, interactive video disc and intelligent tutoring system. Most instructional methods adopted in educational institutions and programmes can be reduced to lecture, discussion, self-learning and distance education. Each method has a large number of variations and each such variation can be combined with one or more variations of other methods while designing classroom instruction. While pure lecture is monoactive, good discussion is interactive and self-learning is intra-active or reflective, distance education is an effective combination of all of them. Experiential learning made possible better by experiential session adopted in the teaching of management sciences is included in self-learning in this scheme of classification of instructional methods. Intelligent tutoring system can be considered in the group of distance education methods.

3.1 Dimensions and Levels of Methods

The three-dimensional three-level approach to describe scholastic subjects can very well be adopted to describe instructional methods also. While lecture is cognitive - dominant, discussion is cognitive - conative interface, self-learning is conative-dominant and distance education adopting individualized instruction is affective-dominant on the basis of which of the three dimensions is mostly taken care of in adopting an instructional method. Variations and combinations of variations would contribute the three levels of simple, average and difficult for conative, simple, moderate and complex for cognitive and direct, covert and angular for affective domains of instructional methods.

3.2 Illustrative Description of Instructional Methods

It would be more useful if we adopt a 7-tier system of lecture, discussion, tutorial, self-learning, workshop, distance education and open learning.

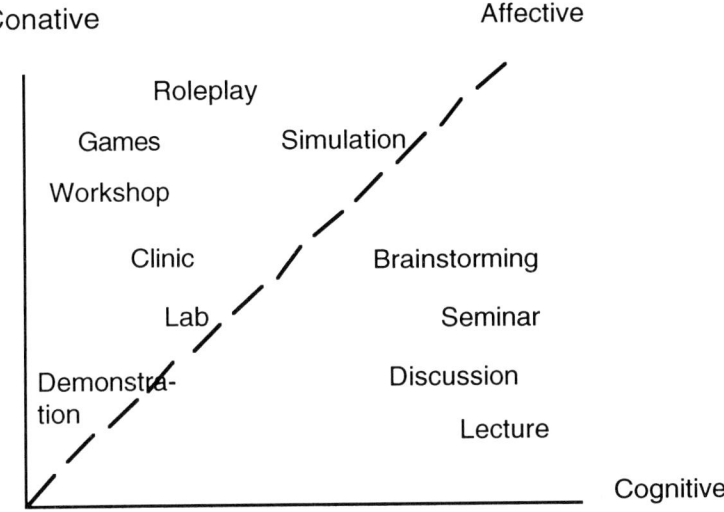

Fig. 5 Instructional Methods

Taking student option for content (cognitive) on X-axis, resources manipulation (conative) on Y-axis and environmental enjoyment (affective) on Z-axis to synchronize with the three dimensions of scholastic subjects, any instructional method can be extricated as a small cube from the total cube. Naturally the position of this cube in the total structure would indicate its level of suitability for teaching the subject. For instance the cubes closest to the zero centre of the graphs of subject domains and levels and of instructional methods indicate the highest relevance existing between that subject and that method.

4 Human Participants in Education

The third component actively involved in the process of education points to human participants, primarily students and teachers, but inclusive of secondary personnel such as parents, educational administrators and even policy-makers. Adopting the same 3-dimensional, 3-level approach to human participants in education the following graph is envolved.

4.1 Dimensions and Levels of Human Participation

Educational participants also fit into the above scheme with their abilities for intellectual excitement (cognitive) on X-axis, for interactive communication (conative) on Y-axis and for interpersonal relations (affective) on Z-axis.

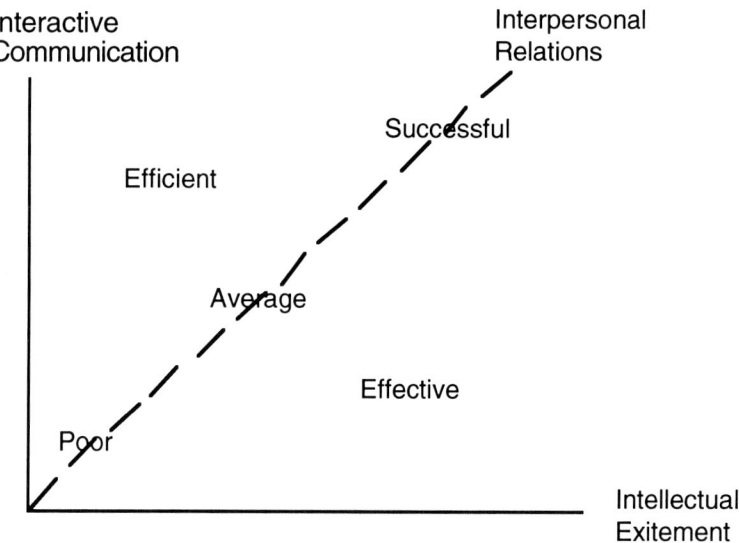

Fig. 6 Characteristics of Humans in Education

If the level in each of the dimensions is closest to the Zero centre it conjoins the description of subjects of study in that cube and the description of methods of instruction in the very same cube. But as human abilities differ the synchronization of methods and subjects with humans points to a range of instructional strategies on the one hand and learning styles and preferences on the other. The place of these styles and strategies in cognitive consonance, and not in cognitive dissonance, shows the optimal success in education.

4.2 Illustrative Description of Specific Human Involvements

In the study of subjects in the cube closest to the zero centre the instructional method identified in that cube would involve individuals in the educational process with the least intellectual excitement, the least interactive communication and the least interpersonal relations. Therefore there is need to make this involvement more dynamic to achieve success with inherent potentials in the individuals. Compensatory mechanism would require changes in styles and strategies adopted in instruction. Similarly in the study of subjects in the cube farthest to the zero centre all the three dimensions operate at the highest level requiring human participants to adopt extreme characteristics to achieve excellence in education.

5 Instructional Design and Learning Preferences

Instructional design of teachers and learning preferences of students weigh very high in the choice of subjects of study and in the adoption of appropriate instructional design to maximise the results. Kanter's (1984) three technical dimensions of information, viz. (i) form, capacity and degree of integration of data; (ii) response time, capacity and interrelationships of data elements, security and validity; and (iii) costs to acquire, maintain and access data, are not only suitable to Management Information Systems but are very well applicable to Instructional Design Development too.

5.1 Superimposed Spheres of Knowledge, Method and Personnel

The superimposition of Personnel characteristics operating with methodological aspects on knowledge structures is essential to educational success and excellence. The analysis of each domain into as many levels as required would help the management of instructional design environment.

6. Implications of the Analysis

The above analysis implies many things to various personnel in education:
- new subjects are emerging with different combinations of levels and dimensions
- curricular formulations are to take care of the three components and their constant movements within this framework
- teacher education programmes have to focus on the outcome of componential analysis of factors acting on each dimension from the angle of knowledge structures, methodological aspects and personnel characteristics.

References

1. Burns, H., Parlett, J. (eds.): Proceedings of the second intelligent tutoring systems research forum. San Antonio, Tx., April 1989
2. Dede, Christoper: Educational and Social Implications. In: Forester (ed.) 1985
3. Duchastel, Philippe C.: Cognitive Design for Instructional Design. In: Instructional Science 19, pp. 437-444, (1990)
4. Flavell, J. H.: Cognitive Development. Prentice-Hall, N.J., 1977
5. Forester, Tom (ed.) The Information Technology Revolution. Oxford, Basil Blackwell, 1985
6. Graesser, Nakamura: The Psychology of Learning and Motivation. New York, Academic Press, 1982

7. Jamal, J. Al-Menayes, Se-wen. Sun: Processing Complex Information: What are the Cognitive units and how are they related? In: GAZETTE 52, pp. 57-84 (1993)
8. Kanter: Management Information Systems. New Delhi, Prentice Hall 1984
9. Martin, Jack: Toward a Cognitive Schemata Theory of Self-instruction in Instructional Science. 13 pp. 159-180 (1984)
10. Nisbet, John, Shucksmith, Janet: Learning strategies. London, Routledge, 1988
11. O'Neill, Jr. H. F., Spielberger, C. D. (ed.) Cognitive and Affective learning strategies. New York, Academic Press (1979)
12. Pirolli, P.: On the art of building: putting a new instructional design into practice. In: Burns and Parlett. (1989)
13. Pirolli, Peter, Russell, Daniel, M.: The Instructional Design Environment: Technology to Support design problem-solving. In: Instructional Science. 19, pp. 121-144 (1990)
14. Sadanandan, P., Chandrasekar, R. (eds.) (1984) Information Technology for Development: Proceedings of the 22nd Annual Convention of the Computer Society of India, February 4-7, 1987, Bombay. New Delhi, Tata McGraw Hill, 1987
15. Winn, William: Some Implications of Cognitive Theory for Instructional Design. In: Instructional Science. 19, pp. 53-69 (1990)
16. Wood, G.: Cognitive Psychology: A Skills Approach. Monterey, Calif., Brokes/Cole, 1983

Psychohygiene in a Manager's Work

Alfred Prigl
College of Zilina
Monzesova c. 20, SK-01026 Zilina, Slovak Republic

Abstract. The author of this paper considers the mechanism of the pathogenesis in the managers' work and provides the analysis of psychologic conditions of effectivity of his productivity. He describes the tools of psychohygiene in his work.

Psychohygiene is the science of studying the conditions of the establishment and maintenance of mental health. Its objective is to give exact reasons for empirical findings.

Keywords. Personality, psychical strain, psychohygiene

1 Psychoreactive States and Some Mechanisms of Their Origin

A manager lives in a difficult situation which comprises a lot of pathogen factors. Some of them are quantitative (high work load), others are qualitative (family problems, strain, interpersonal relations in the work place, etc.)

Pathogenesis (i.e., the origin and development of pathogenic states) results in diseases or states which are called psychoreactive (i.e., conditioned by psychical influences).

Psychical life can exist due to its biological presupposition - the brain. We can simply say that the brain is divided into two systems: specific and non-specific. The anatomical and physiological divisions correspond to two types of psychical activities:

1. non-specific, or general; a little-directed activity of general, preparatory, ('alarm') character.
2. specific; a precisely directed activity which reacts to stimuli discriminatively.

When exposed to an unexpected stimulus, the living organism comes into a state of 'alarm', muscular tonus increases, vegetative reactions, which prepare the organism for an eventual danger, appear.

The first phase corresponds to the activity of the first system, the second phase, when a conscious analysis is being done, is the expression of the second system. The first, the so called alert system, is responsible for psychical strain which is a subjective reflection of the readiness of the organism. Psychical strain as a state of readiness, activation, influences both psychical and motor performance and activity in general.

The relation between psychical strain and psychical performance and their mutual dependence can be expressed graphically by means of two curves:

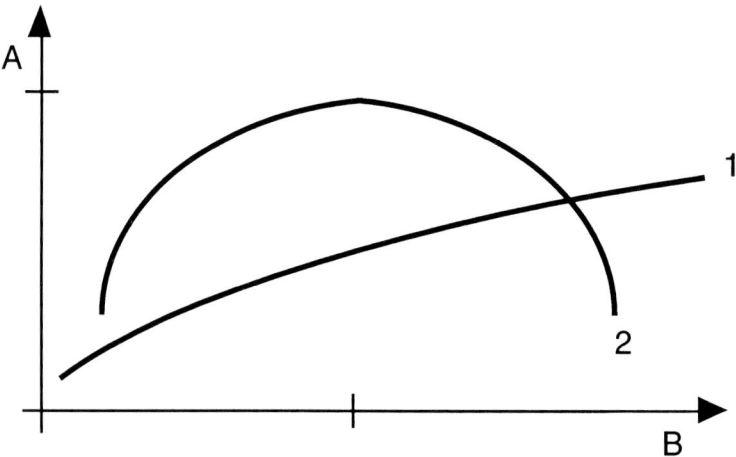

1. a relation of a simple performance
2. a relation of complicated structured performances (the law of the inverted U)

Fig. 1 Relation between psychical strain and psychical performance

Psychical strain improves performance in a monotonous way, in case of simple, little-structured performances. For complicated tasks, i.e., tasks requiring sophisticated, skilled or creative ways of solutions, performances depend on a strain in compliance with the law of the inverted U. The optimal performance is at an intermediate level of psychical strain. During supraoptimal strains, performance decreases. For example, if I am to solve a complicated scientific task and I am nervous, I can be absolutely unable to carry out performance.

The very necessity to solve a task creates psychical strain. If I do not solve a task, strain increases. We speak of a feedback regulation. If I am unable to solve a task adequately, (e.g., under a high psychical strain or my problem-solving competence is not sufficient enough), in a slightly differentiated way, an increasing strain requires any solution: I will, therefore, solve a task in a way which is a little differentiated, i.e., regressive. (In principle, there are only two types: aggression and escape, both having an unlimited number of forms). Under high psychical strain there is a regression of solutions to a lower level of complexity.

In this way dissociation appears, i.e., the separation of the functions of two brain systems. During a complicated task, the non-specific system of the brain evaluates the situation as important and comes to a state of a high activation - alertness. The state of a high activation prevents the specific system from a fine analysis of the situation and from a precise solution. Primitive, simplified, stereotyped and regressive solutions are preferred.

Dissociation can be often seen in everyday nervousness. The cause of this failure is not quite simple. There is a long sequence between cause and effect in a psychical life. Some other phenomena appear. They are:

– Mechanisms of a personality
– Frustration
– Motivational disparity
– Conditioning.

1.1 Mechanisms of a Personality

The brain activity is highly automated and the basic brain integrations are subconscious, (e.g., I am unaware of the fact that the brain controls my muscular strain). A conscious, subjectively experienced activity appears only at the highest level of integration.

A subconscious integration controls the activities of individual functions, e.g., motional, secretionary, etc. There are also subconscious operations which control an activity of a personality as a whole. They are called mechanisms of a personality.

They are operations common to all people. They are similar to inborn reflexes. They operate automatically, they come into operation involuntarily and without any immediate conscious control. They are simpler than an exactly discriminated conscious activity. In general, it can be said that mechanisms of a personality operate when a human being cannot consciously control his activity. They are expressions of a state of psychical crisis.

Escape or aggression are frequent mechanisms of a personality. An escape is an extraconscious avoidance of the situation I am unable to solve, (e.g., I feel like having a drink, a fantasy to illness, etc.). During aggression I start a conflict in order to avoid a solution.

The mechanism of a personality is a regressive solution. It suits a dissociation activity scheme of two types of psyche we have already mentioned. In a complicated situation there can be a high psychical strain which prevents (according to the law of the inverted U) from an accurate conscious solution. The mechanism of a personality can appear as a supplementary solution, as a subconscious activity which cannot be analysed rationally. (The mechanisms of a personality result in a chain propagation of regressive phenomena.)

1.2 Frustration

An organism has certain needs. Some of the needs have roots in the biological basis - biological needs (eat, sleep, breathe), others in the social basis (selfrealization, certainty, appreciation, etc.). The need has a specific, usually rhythmical way of feeding. If the feeding is somehow blocked, an unpleasant situation - frustration - appears. Psychical strain automatically increases during frustration.

The primary frustration is a feeling of a specific insufficiency as a demonstration of frustration (I am positively aware of what I miss). If frustration 'propagates', it is manifested in a total annoyance, sulkiness. It is the secondary frustration (e.g., hunger, sleeplessness. I am unaware of the cause.).

When some social needs are insufficiently fed, almost in all cases the secondary frustration is developed. The primary frustration need not to be conscious at all. In such a case the primary frustration can be incompatible with a real self-assessment. The secondary frustration is a source of a psychical strain, which can deteriorate a work performance. In the secondary frustration the cause of strain can be quite far from the situation in which it manifests (e.g., erotic). Due to frustration and a subsequent rationalization, a personality experiences his/her situation in a distorted way, does not understand himself/herself, designs incorrect versions.

1.3 Motivational Disparity

It is another important source of strain. It is a disparity (disharmony of motivations). Motivation is always presented in human activities as a matrix having both positive and negative values. Therefore, instead of a monosemantic motivation there is a polysemantic and conflicting motivation (collision of interests). In general, a conflict is the source of strain. The strongest is the collision of interests, the highest is the psychical strain.

1.4 Conditioning

In psychoreactive states there are many manifestations which can be derived from the theory of a conditioned reflex. Conditioned reflexes extinguish inproportionally. Extinction is an elimination of a conditioned response when only a conditioned stimulus is repeated several times without being reinforced by an inborn stimulus. Schizogenesis takes place. A specific component of a conditioned response, a directed movement, extinguishes, but a non-specific, vegetative component lasts very long. A specific component of the conditioning has extinguished, a non-specific component has remained.

Many neurotic manifestations can be considered as non-adaptive, persistent fragments of conditioned responses which were some time ago useful. The fragments gradually lose their original meanings and for the human being, who is their bearer, can be quite obscure.

The mechanisms, we have talked about, are mutually interrelated and control the feedback directed circuits which, sometimes, overlap the boundaries of a personality. E.g., a person 'A' showing an increased physical strain can infect his/her partner in the marriage. Psychical strain is infectious. In this way, not only the performance, or state of the original person 'A' is influenced, but also the partner 'B'. The person 'A' can rationalize his/her unpleasant situation so that he/she accuses the partner 'B' of having developed the situation and vice versa. Such crossed rationalizations are frequent. Both partners spread their strain further - to their work places. In this way, some other people are trapped into anomalous feedback regulations. A system of vicious circles develops and gradually expands. It is, in fact, an epidemiological problem. A poorly adaptable person infects his/her environment. The genesis of this network consisting of circles is irrational and the participants need not be necessarily aware of the true causative relations.

2 Psychohygienic Consequences and Prevention

In the previous sections we have explained why and under which circumstances psychopathogenesis operates. Now there is another question: what mechanisms protect a person from psychopathogenesis and why does not everybody become a victim of psychopathogenesis.

Apart from non-specific mechanisms of the brain activation, new, more specific mechanisms are being developed. They ensure an elective and specific act of regulation. These specific performances can be seen in three areas:

1. symbolization, i.e., the development of meanings from originally meaningless objects.
2. autoreflection, i.e., an ability to see one's own inner state as something 'external.' Autoreflection enables a person to deactualize his/her emotion - to make it less urgent. (E.g., when I can see myself as somebody else, or from a bird's eye view, I do not experience my misfortune as eminent, I am able to 'uplift' myself). A significant way of autoreflection is self-irony, which can have a relieving effect (if I am able to make fun of myself, I shall not suffer so much as if I take everything seriously).
3. frustration tolerance - is an ability to acquire a partial immunity of a new frustration on the basis of experienced and overcome frustrations.

Both, symbolization and autoreflection enable a quite unique human ability, an attitude to extrapersonal, or above-personal values. A human being synthesizes the relations which are not his/her personal property, if they have a character of extrapersonal regulating principles (e.g., truthfulness, justice, etc.). The values are of a complex character: on the one hand, they must be developed by the human brain, on the other hand, they belong to the whole society, they do not belong to a single person.

The identification with a value can significantly regulate human behaviour and personal experience, e.g., I can be personally very unhappy, but getting involved in a solution of a scientific problem which is of great value for me, makes me happy and I do not pay attention to my misfortune, (it is a kind of 'happiness on the second floor').

Autoreflection decreases psychical strain. In compliance with the law of the inverted U, it enables to influence adequately the performance in a complicated situation. A systematic confrontation with extrapersonal values, autoreflection and auto-criticism have, thus, curative and preventive effects. It means that psychohygiene is directly related to philosophical and ethical problems.

From the psychohygienic point of view, it is important for a human being to develop the system of high-quality values - a correct value orientation serving him/her like a compass to guide him/her in complicated everyday situations and helps him/her in an autoreflexive way.

The system of individual values should be permanently critically re-assessed. They should be sufficiently differentiated. The individual 'monocultural' value can be harmful. The orientation to one culture can be catastrophic if the value fails (if a violinist lives for music only and he loses his hand). The system of individual values should have a certain plasticity and elasticity.

The monoculture of values can be disadvantageous also from the point of view of creativity. If there is an intensive concentration on one solution of a problem, possible solutions of a problem are narrowed from more aspects.

Strain and anxiety make a psychical space smaller. Decrease in strain, deactualization of an emotional state, psychical comfort offer more possible solutions. From the point of view of psychohygiene, two conclusions are important:

1. A manager should have values not related to his main occupation (i.e., a garden, sports club, stamp-collecting, fishing, etc.). In this way there are more possibilities for his autoreflection and distantness.
2. In his life-style there should be a possibility for 'creative indolence', i.e., a purposeless and involuntary correction and re-make of his ideas.

The approach to the values and selfreflection do not operate as single acts, but they exist as a systematic style. There should be a proper education provided. Self education is a starting point for psychohygiene. The systematicness is inevitable as it has already been shown, when dealing with conditioning. Drill can develop such behaviour and attitudes which cannot be achieved by a single act.

Some forms of psychical strain and their consequences cannot be influenced directly. Indirect methods have to be used. Psychical strain can be suitably influenced when the muscular strain is influenced. Both muscular and psychical strains are closely related. The increase of psychical strain is always reflected in an increase of the muscular strain, either in general, or in certain muscular groups. e.g., a headache during 'nervousness' is often induced by the increased strain in the muscles in the areas of the neck and forehead. On the contrary, the decrease in the muscular strain, always and reflexively, i.e., automatically, evokes the decrease in psychical strain. If we succeed in lowering the muscular strain when we get excited, we automatically keep our anger down. The feedback relations between muscular and psychical strains can be depicted in this way:

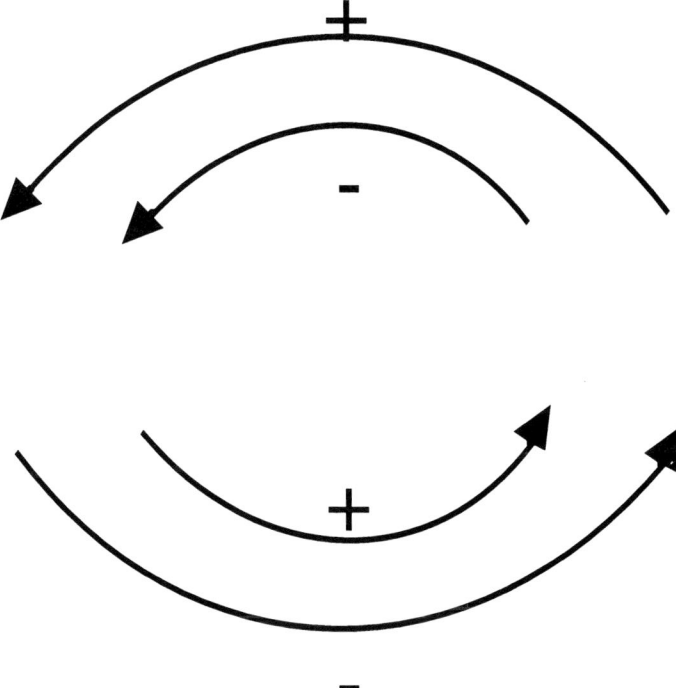

Fig. 2 Psychical strain muscular strain

This knowledge is the basis of various relaxation techniques. Their objective is to teach a human being to govern his/her muscular strain and thus, he/she is able to control his/her psychical strain in an indirect way.

Relaxation can exist not only as the means eliminating directly pathogenous regulation circuits (as it improves a possibility for a finely differentiated solution of problems), but also as the means of conditioning.

There are techniques which eliminate fixed improper reactions. They are based on the principle of reciprocal inhibition (reflexive restrain). According to this very general physiological principle, the development of a reaction automatically inhibits the reaction which is biologically incompatible, (e.g., the activation of the breathing-in centre automatically inhibits the activity of the breathing-out centre and vice versa.). The mechanisms of psychical strain are also governed by this principle. If such a situation which is incompatible with psychical strain is developed, it is automatically suppressed. There are two states of this kind:

a) relaxation - a direct antagonism of strain
b) self-assertion, i.e., an offensive attitude. Excessive psychical strain is a defensive state from the biological view-point, it is an expression of impotence of the organism. Self-assertion is biologically incompatible with excessive strain.

These procedures are applied in artificially constructed processes, but a human being makes use of them in a natural way.

Included in it is also adequate body load. The purpose is not only to increase the physical conditions, but also a purposeless development of self-assertive training. Also during the relaxation which is a result of autoreflection, when experiencing, for example, an artistic performance, we can simultaneously get desensitized from the stimuli which were infectious, (e.g., imaginations evoking unsuitable conditioned responses).

In the end it has to be seen that all the processes we presented here, have their material substratum, the brain, which has certain genetic characteristics. These characteristics can be specified by means of the upbringing, environment, etc. There are persons who are stable, regulatively 'qualitative', others are more labile with the regulations which are easily put off.

The physical substratum can be damaged also by the acquired influences having a character of physical functions of different kinds. Psychohygiene has to pay attention also to them.

Most of them are well-known. It is abuse of alcohol, drugs, too much coffee, lack of movement, a chaotic rhythm of a daily programme. They impair the brain performance in general. If, e.g., the concentration decreases due to drugs, there is an automatic increase in the psychical strain in a problematic situation and the pathological regulation comes into operation.

It should be obvious that psychohygiene cannot be a guide how to behave in a certain, predetermined way. It has to respect the personal uniqueness. It can be an aid for self-knowledge from a certain aspect and also for a construction of a systematically reflected life-style. In this point, it is closely connected with other sciences, the objective of which is to shape human behaviour. It is, at the same time, the means of 'an equipment' necessary for every manager.

Developing a Competency to Act Within Networked Systems

Walter E. Theuerkauf
University of Hildesheim
Institute of Applied Electrical Engineering and Technology Education
Kreuzstraße 8, D-31134 Hildesheim, Germany

Andreas Weiner
University of Hannover
Institute of Ergonomics and Didactics of Mechanical Engineering
Im Moore 11A, D-30167 Hannover, Germany

> *"Technology's great, but we believe in relying on human judgement, not machines."*
> *Donaldson 1994*[1]

Abstract. The paper deals with a novelty – teaching plant – a model production system as one means of technology education. Sub-sections are networked for optimization of material and information flow. The team conference is a part of the group work – another form of cooperation. Optimum plant performance as an evidence of the performance in the learning process.

Keywords. Computer control, course concept, curricular objectives, networked factory structure, teaching plant

1 Introduction

The production patterns and organizational structures in the automobile industry has undergone profound changes within recent years (cf. Stotko 1992). These changes are to a very large degree characterized by new strategies of personnel

---

[1]William Donaldson: Chairman of the New York Stock Exchange, who is trying to stimulate a new way of thinking in realizing stock exchange transactions.

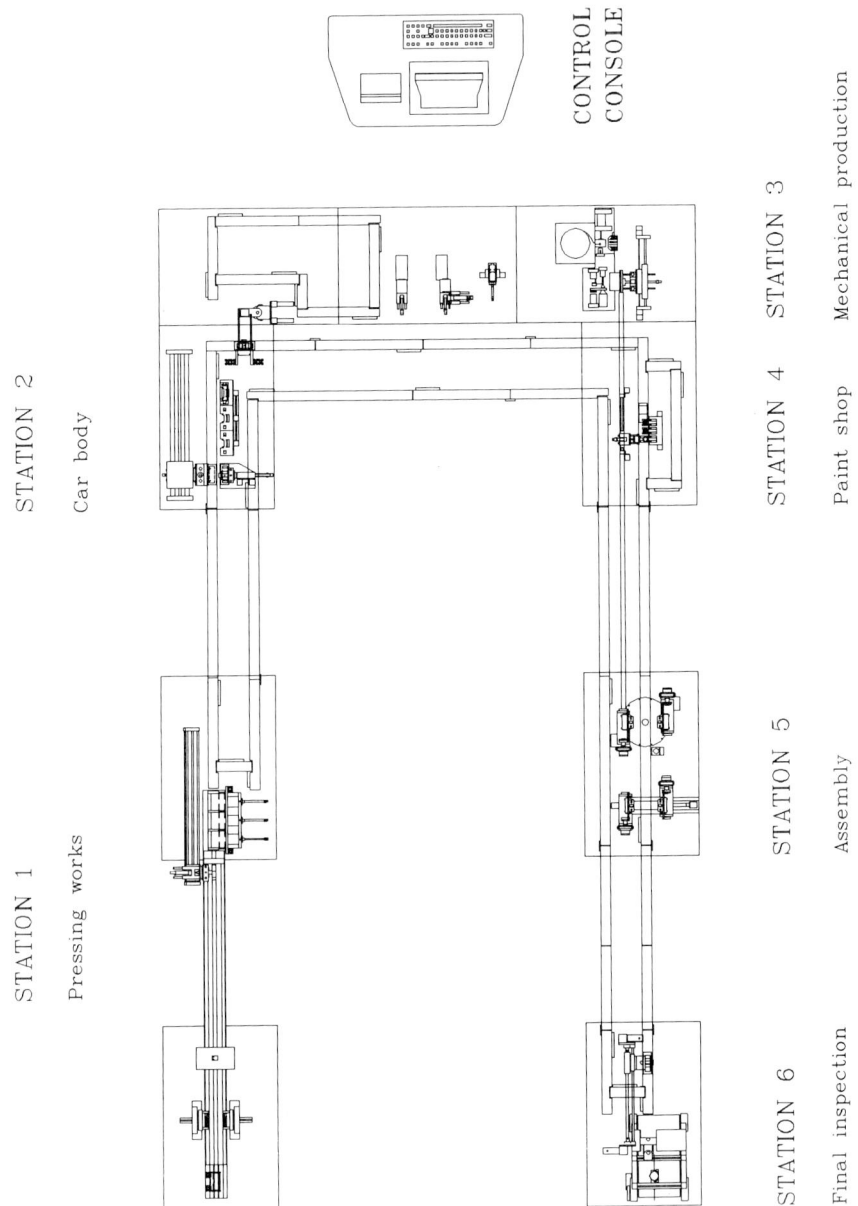

Fig. 1 Layout of the model production system

management that go along with a new employee profile. If business objectives are to be achieved, the consequence has to be a new quality of staff qualification.

The networked factory structures in companies have the following consequences for the communication of expert, methodological and social competencies:

- the development of process competencies, allowing employees to handle complex sequences of operations successfully and place them in the proper perspective within the general context of the company;

- the development of methodological competencies to be able to master networked operations involving directive and optimizing functions;

- the development of social competencies and compentencies in work and technical organization allowing employees to arrive at the appropriate task-oriented decisions within the group and to accept, as an individual and as a team, responsibility for the production process (Theuerkauf und Weiner, 1991).

It will for the sector of in-company training and education be demonstrated below in what way the key qualifications listed above can be developed with the aid of a networked production system where a model car is produced.

2 System Configuration for the Model Production System

To be able to develop company-specific key qualifications, a model production system[2] was developed serving as a teaching plant. This system provides for the production of 54 versions of one model car thus copying the product range of the company for whose training centre the qualification concept had been developed.
The teaching plant is an assembly plant comprising six self-reliant but networked learner subsections, with on the one hand a material flow that is determined by the structure of production cells (micro view), while on the other hand there is the plant that represents the product-related production of a model car (Fig. 1). The material flow between the sections follows a linear pattern; including buffers. The sections represent different production sectors of the company, as for instance the press or the paint shop.

Production planning and production control proceed from the control centre[3]. To be able to simulate the sequence of the production steps, commencing with the vehicle as ordered by the customer, the production planning system depicted in the control centre followed the normal company practice and included elements of the forecasting phase and of production planning. The company's very complex programme packages with the associated tasks were reduced to the essential actions for a TOP DOWN method.

The interfaces between man and production equipment are tied to the actions operating, running and supervising the sections. The operating panels and the electric control circuitry were designed with due regard to company standards and safety regulations. With the aid of monitors the material flow can be followed and watched, actuators can be manipulated, and signal conditions of sensors can be interrogated. Course participants responsible for operation of the sections are left

---

[2]Responsible for the design and technical realization of the model production system are VW AG Wolfsburg and FESTO DIDACTIC KG, Esslingen/Germany.

with a number of possibilities of how to manipulate the function of the sections in the modes "adjusting", "individual operation", "individual movement", and to follow the process within the networked system.

The operator's ability to respond quickly and adequately to faults within automated systems is an essential productivity factor for real production systems. The course therefore provides for a training programme where under an interlinked operation scheme faults have to be eliminated. To this effect, different fault configurations can be combined or faults can immediately be triggered in the control centre. The master computer initiates mechanical, pneumatical, electrical or EDP-related faults, which is either a random or an immediate, but always a section-related process.

When considering the different levels of the company structure, model production suggests the following actions on the different levels:

Planning level: production planning, time scheduling, etc.;

Cell level: production control, progress control, SPC;

Process level: resource control.

The model production system is thus available in the form of a group-based laboratory, which on the one hand simulates the actions within the company structure and on the other hand serves as an educational system. But only by cross-linking the individual learner sections can the development of key qualifications actually be achieved.

With a view to information theory, the model production system is a medium that possesses a stimulative character and is because of its complexity at the same an information store as reflected by its complex system/process functions. The model production system thus represents a novel educational dimension for the learning process, which is based on a new learner/medium relationship. The complexity of the task allows - in conjunction with the model plant - to acquire the capacity to act on the different action and control levels.

3 Realization of the Curricular Objectives

Adequate qualification analyses are an indispensable prerequisite when establishing a catalogue of educational objectives for a course. But two general objectives can be named irrespective of such analyses.

- In the course, the group is to aim at an optimized production performance within the model production system. To achieve this, special emphasis is placed on the demonstration and the use of restarting, run-up and continuous-operation strategies (cf. Millberg et al., 1992).
- To maintain a smooth material flow within the production process of a model car, operators have to be able to act competently and reliably. As the model plant comprises six networked sub-sections, optimization of the production process

cannot be realized by actions of the individual, but only by the group becoming active in the learning process (Bühner und Pharao, 1993). The linear structure implies a retroactive effect between group and individual, thus also necessitating individual commitment.

The objective of trouble-free continuous operation providing at the same time for acceptable productivity standards can only be achieved when strategies of thinking within networked systems are available. To acquire the required expert competency, the course concept concentrates on the key issues listed in Table 1.

Table 1 Factors of the expert competency as part of the courseconcept

Production strategies	Knowledge of the systems and the structures of logistics in the automobile production sector. Drawing a comparison between the company production system and the model production system. Realizing and identifying evaluation criteria for production and assessing same.
Programmable controllers	Knowledge of the configuration and programme of programmable controllers and the way they are linked with EDP systems. Knowledge of the mechanical, electrical and pneumatic elements of the system. Identify functional structures of the controllers, i.e., analyse the logic circuits between the mechanical, electrical and the automation units. Identify and eliminate faults.
Communication/Inter-linking of resources	Knowledge of the in-company networks and the network available in the system, including data acquisition, data transfer, data manipulation, data evaluation. Describe the structure and function of the information flow and identify their effects on the energy and material flow.
Optimization of material	Knowledge of the strategic, operative and information flows objectives of the company. Ways and means of optimizing the material flow within the model production system. Bring measures of organization, qualification etc. into a relationship with the model production process and translate same to actual working conditions in the company.

As a methodological competency, the ability to act competently and reliably necessitates the ability to inform, think, decide and react independently. The key qualifications this requires are listed in Table 2.

Table 2 Factors of the methodological competency as part of the course concept

Analytic, step-by-step thinking	Split up the functions of the different sections of the entire production process into logic sequences with the aid of the system analysis, which is then represented and discussed within the group.
Synthetic/conceptive thinking	Develop production and qualification schedules. Influence the daily production.
Acquiring and processing inform.	Analyse and evaluate process information furnished by section monitors. Analyse production results. Develop alternative production strategies. Consult circuit diagrams and manuals for fault analysis and elimination.
Methodical approach and selection of information	Working with the Leittext (or keysmaterial) method. Selection of the necessary elements of knowledge providing for better production process handling.
Problem solving	Identify malfunctions within the production process, i.e., bottlenecks, and develop solutions for process optimization. Employ problem solution strategies also as trouble shooting strategies and eliminate faults.
Transfer	Contribute industrial experience and newly acquired knowledge to the learning process and translate same into practical action in start-up and control procedures within the model production process.

Commitment and successful action within the group presupposes an adequate social competency. The relationship between this competency and the actions within the production process is demonstrated in Table 3.

3 Methodological Features of the Course Concept

If learning is to have the effect of developing professional competencies, a course trying to communicate such competencies should aim at aspects that will be of significance for the target group in its real job situation.
Starting from this consideration, the course for system supervisors was conceived such that the model production plant was to be put into operation and run in at steps of 25%, 50% and 100% of its maximum performance (Table 4).

Table 3 Factors of the social competency as part of the course concept

Flexibility	Assume different tasks in the production process and adjust to new situations (e.g., disturbances) and functions within the learning process.
Capacity for teamwork	Master the model production process and the course situation as a production team/group of learners both by dividing the work and by a common approach. This also presupposes that the role of the trainer/moderator is that of a partner.
Ability to communicate	Come to an agreement between the and negotiate subteams responsible for the different sections. Maintain communication with the trainer. State and evaluate facts. Oral and written presentation of problems.
Initiative	Develop solutions to problems occurring in the production process without having been asked to do so by the trainer; discuss possible solutions with the group and translate them into action as agreed with the trainer. Suggest learning processes on the basis of an analysis of individual qualification deficits and initiate same individually.
Responsibility	The production plant represents an asset demanding careful handling. Improper handling has consequences for the work/the learning process of the present and of future groups.

The course thus follows normal run-up procedures in real plants (cf. Wiendahl, und Garlich, 1992) and hence has immediate reference to the responsibilities of system supervisors.

A prerequisite of qualified group work on the job is the core task (Gohde und Kötter, 1990). The core task of this qualification concept and hence a point of reference for the group is the achievement of the optimum model plant performance or output. Derived from this core task are sub-tasks determining the individual learning steps to be acquired. They are developed with the group in team conferences and geared to the given objectives of the qualification plan. Such an approach demands from the learning group not only creativity and innovative abilities for the work to be accomplished, but also for the individual learning process.

Table 4 Typical course structure

Exercise	Method/Media	Teaching Operations
Phase 1: – Capability for individual learning – Strategies for starting the plant – Strategies for problem solving	– Combined instruction – Lecture by the coach – Guide text – CBT/Video	– Working with guide text to develop key qualification

Exercise	Method/Media	Teaching Operations
Phase 2: – Planning the initial and the starting operation – Initial operation of some fundamental units	– Grif-plant – Self-learning with starting algorithm and documents of the plant	– Initial operation under the guidance of the coach – Analysing the defects during the starting run
Phase 3: – Job Order manufactoring with the capacity of 25% – Training at the different stations	– Master control – Grif-plant	– Starting the networked stations – Production control – Analysing the results of the production – Discussing the statistics of the production
Phase 4: – Job Order manufactoring with the capacity of 50% – Recognizing and eliminating defects – Improving the production flow	– Master control – Grif-plant – Fault-generator	– Production control – Analysing the results of the production – Discussing the statistics of the production
Phase 5: – Job Order manufactoring with the capacity of 100% – Improving the production flow	– Master control – Grif-plant – DV assisted analysis of the weak points	– Production control – Analysing the results of the production – Discussing the statistics of the production – Pursueing the targets of the production

Phase 6: – Individual and collective analysis of the sequence of the course	– Analysis of the weak points of the plant	– Individual planning of further education

Collateral course objectives are trouble shooting and fault elimination during different production phases. To be able to successfully practise trouble shooting procedures, the strategies employed start from the theory of problem solution and correspond to the diagnosis procedures taught in in-company training programmes (cf. Genath 1984). They were included in the course in order to reduce to a minimum plant stoppages and downtimes during the running-up phase.

Familiarization of the participants with the start-up procedure in the different plant sections is not an instruction item. From their work as system supervisors, most participants already know the configuration of the operator's panel as well as the algorithms to be able to start the plant from the panel. What they normally do not know is the structure of the monitors of the programming unit and the way how the monitors are linked with the different operating conditions of the production system. The participants acquire this knowledge autonomously from manuals studied in parallel with the practical approach.

A significant aspect is that in this way the more selective or specific knowledge participants have acquired through experience is linked with newly acquired knowledge. The experience and abilities of the participants is used to develop a dynamic working team where group-oriented problem solutions are given priority. Hence the course structure will not be the same for other target groups.

The qualification concept making use of the model production system provides in a very special way for the development of an individual behaviour (Bunk, Kaiser und Zedler, 1991, p. 369). For the development and furtherance of a social behaviour, such as the ability to cooperate and communicate, the course employs social patterns permitting the individual to work and learn with his or her colleague or within the group as a whole.

Another form of cooperation is the team conference, the main issue of this conference being the starting of the production process, increase in production and the productivity of the model production system. The team conference encourages:

- the shaping of the action pattern required for plant start-up by describing concrete objectives and sub-steps, as well as the accepting of a shared responsibility for the learning process
- arrangements on the functions to be performed by the participants for plant operation, to name but two aspects.

In the team conference, participants do not only accept responsibility for the production process, but also for the learning process. Before commencing with each item to be acquired, the participants discuss all the tasks that have to be completed or that have already been solved. In this context the trainer is assigned the role of a moderator.

5 Productivity Considerations as a Benchmark for Progress

If the aim of the course is to arrive at an optimum plant performance and output, and if the production statistics represent the result (controlling) of this process, they can also serve as an evidence of the performance in the learning process. They comprise the practical abilities in a comprehensive way as well as the theoretical standard of knowledge. The advantage of this progress monitoring lies in the objectivity of the evaluation method. Since each course covers at least three production cycles, the production statistics they produce are direct indicators of the knowledge gained (Fig. 2). The statistical characteristics are interpreted as follows:

Table 5

Capacity utilization	shows the level of the production performance of the group as compared with the plant rating,
Running time factor	shows the downtimes for the plant sections,
Degree of utilization	shows in how far the objectives of the group (intended car output) have been achieved,
Error statistics	show the nature, instances, significance and times of trouble shooting/elimination processes.

Statistical data can also be compared with the results of earlier courses and used as a basis to discuss improvements.

Such an approach in assaying progress provides course members with an economical and practicable instrument of self-control without having to force the trainer into a role of a critic or judge.

6 Final Remarks

The concept presented here was subjected to a summative evaluation in a number of pilot courses. It was proved that the intended objectives can be met with the model production system.

The model production system provides for active non-simulative learning by doing for the actual job. For the learner it proves to be an experience- and perception-based method of presenting and activating existing knowledge. It was found to be important to develop the ability to act reliably in systems characterized by a high degree of automation. The fact that the learners did often

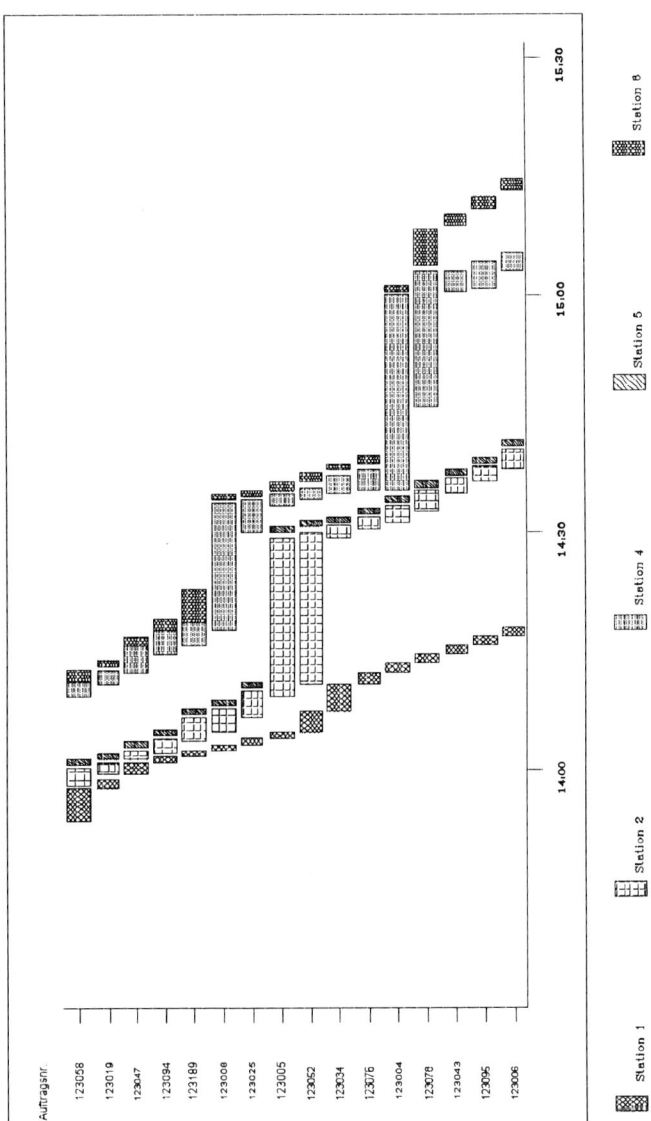

Fig. 2 Status of the Job Orders

not achieve the intended production result within the given time demonstrates the very high demands complex networked systems make on them. The concern especially of system supervisors showed when failing to achieve the maximum output, which could be explained by their professional self-image, proved to be an encouragement for a yet increased commitment. The controlling proved to be conducive as statistics demonstrated quite clearly whether the production cycle concerned provided for a real net output or not.

For the company, the 'model production' laboratory with its synergistic approach will offer a new dimension of experiencing the computer-integrated and networked production process, as in a 'learning' company it can serve as a tool for a learning process where experimenting and discovering figure large. Networked systems assume integrative functions as they call for a coordination of competencies, as well as a group-forming behaviour to allow capabilities to be fully developed. The concept of networked systems is based on a discoursive group consensus and hence it would be expedient to jointly qualify persons on the plant who perform different tasks and who come from different levels within the company hierarchy. This would facilitate not only an exchange of expert knowledge, but also the development of the communicative competence. With such a communication- and relation-based discourse, networked systems comply with an essential quality feature of the learning organization (Grosse, 1994).

References

1. Bunk, G. P., Kaiser, M., Zedler, R.: Schlüsselqualifikationen - Intention, Modifikation und Realisation in der beruflichen Aus- und Weiterbildung. Mitteilungen aus der Arbeitsmarkt- und Berufsforschung (2), pp. 365-374 (1991)
2. Bühner, R., Pharao, I.: Erfolgsfaktoren integrierter Gruppenarbeit. VDI-Zeitschrift 135 1/2, pp. 46-57 (1993)
3. Genath, H.: Diagnosetraining - Eine Einführung für Mitarbeiter der Instandhaltung. Internal script of the VW training programme, 1984
4. Gohde, H.-E., Kötter, W.: Gruppenarbeit in Fertigungsinseln. Nur Schönheitsfehler oder mehr? Technische Rundschau 82/44, pp. 66-69 (1990)
5. Grosse, W.: Fit für den Wettbewerb - Organizational Learning. Q-Magazin 1, pp. 10-12 (1994)
6. Millberg, J., Aman, W., Raith, P.: Beschleunigte Inbetriebnahme von Produktionsanlagen durch getestete Ablaufvorschriften. VDI-Zeitschrift 134 (2), pp. 32-37 (1992)
7. Stotko, E. C.: Automobilindustrie im Umbruch. VDI-Zeitschrift 134 (2), pp. 112-117 (1992)
8. Theuerkauf, W. E., Weiner, A.: Die Vermittlung von Schlüsselqualifikationen mit integrierten Lernorten, dargestellt am Beispiel des flexiblen Lernlaborsystems (FLS). In: CIM Qualifizierung in Europa. Tagungsband. Institut Technik und Bildung, Universität Bremen (Hrsg) Bremen, S. 61-72 (1991)
9. Wiendahl, H.-P., Garlichs, R.: Unterstützung der Inbetriebnahme und des Hochlaufs komplexer Montagesysteme durch ein datenbankgestütztes Hochlaufmanagementsystem. Internal report of the DFG project "Inbetriebnahme komplexer Montageanlagen". Hannover, 1994

New Approaches Towards Interdisciplinary Technological Enlightenment

Ingrid Lisop
Association for the Promotion of Work Oriented Research and Education
Am Eschbachtal 50, D-60437 Frankfurt/Main, Germany

Abstract. Within the Management Movement the so-called New Thinking has been discussed and practiced for almost a full decade. It is focused on the category of 'entirety' which although-not identical with new holism and the network problem, is related to both of them. If we want to adopt new approaches towards technological enlightenment, we will have to redefine the corresponding unities of parts and whole of economy, ecology and technology and relate them to one another for curricular requirements.

For that purpose theoretical instruments as well as practical examples are introduced in the lecture. They are also meant to correlate the tasks of the management on the one hand with those of the engineers on the other hand in an integral way.

Keywords. Technological enlightenment, management, technology

"Enlightenment is an adventure, qualification a reliable thing."

This sentence which you can often hear from politicians was recently proven in a session of the German Bundestag, the legislative assembly of the Federal Republic of Germany. The following was reportedly said, "The time will soon come when in the event of a bursting waterpipe sufficiently qualified university graduates will be able to calculate the increases of water level in your living room, but hardly anyone will be there you might be able to fix the pipe".

This sentence was quoted by a Christian Social Union's member of parliament. Originally, it was coined by the chairman of a well-known Junior Trade Circle. The example shows what the struggle for and against education for and against enlightenment basically is all about, and I deduct my first thesis from it:

1. In the struggle for or against education, qualification or enlightenment economy and social hierarchy are always involved. In our case the worries of medium-sized

industry and of the skilled trades about finding young talents are as much at stake as the fight for privileges. I am going to illustrate this last point as well.

The debate in the Bundestag I quoted from dealt with the report of an inquiry commission which had been charged with the research on fundamental requirements of a future educational policy, in particular with regard to vocational training and university education and in special consideration to the technological changes. The member of parliament I already mentioned, furthermore quoted a little passage from one of my books. Here is what it says, "Finally: Did the one who passed a motor vehicle mechanics training with an honours' degree get any understanding along with it on the wise or senseless use of motorcars? Has he become competent to co-operate in safeguarding our cities from further devastation and our natural environment from further destruction? Has he even learnt to take an interest in it?" The honourable member of parliament commented on these sentences as follows, "Here is my personal opinion: I need a motorcar mechanic who is able to fix my car and I don't want anyone who is only discussing traffic policy and environmental questions with me. I say this quite clearly." I shall evolve two further theses for my subject out of this example:

2. Enlightenment touches on the polarization of democrats and expertocrats or, more generally speaking, democracy and expertocracy, a polarization which is already established in all industrialized countries.

3. Enlightenment is only feasible, at least partly, through a holistic or dialectic approach which avoids thinking in either-or-categories. This makes it so difficult to handle in modern scientific practice and so explosive from a political point of view.

I would now like to explain this thesis. Educational policy means social policy. The German Bundestag has remitted the interim report of the inquiry commission on future educational policy to the following committees: foreign and domestic affairs; economics; labour and social order; youth; family, women and health affairs; technology and assessment of technological impacts; economical co-operation (development aid); environment, nature protection and reactor safety! If this range of responsibilities is typical for industrialized countries education is one of the main subjects of policy - though in a non-direct way.

It looks as if we were facing a big dilemma. On one hand we need experts; on the other hand we need more creativity, more democracy which not only should be able to support innovations but also to put them into practice and cushion them socially. The era of encyclopedias is over, once and for all, and we need 'entirety' yet. How can we proceed?

First of all I would like to show you the tendencies in central Europe in the field of skilled workers' and similarly skilled staff's training and which innovations already do exist in this field. Then I will try to transfer the results to the level of university education.

For a number of years the International Labour Office in Geneva has been working on the development of a system of training modules for international application. The so-called 'modules of employable skills' - there are nearly 1000 in the meantime - are training units which are within reach of everybody, irrespective of the individual educational background. Apart from this development and ever since the great controversies about automation first began in the fifties, there are demands to more strongly emphasize general education and the development of mind, often called 'key-qualifications'. In countries with a longer democratic tradition these demands are aimed at autonomy, tolerance, willingness to assume responsibility, communication, co-operation. However, in countries where the democratic tradition has been interrupted we can observe how the target of personality development is subordinated to operational logistics: qualifications which are demanded there are the abilities of passing information, planning, decision making, executing, controlling and evaluating. By the way, I wonder whether the much-cited success of the German system of vocational education is less based on the combination of vocational school and acquiring practical skills in the manufacturing plants, than on learning to adapt to the requirements of operational logistics and to the key-qualifications.

Well, beside these two trends towards modularization of training and emphasizing the development of mind, a third one may be defined, i.e., the polytechnical training which means the imparting of polyvalent qualifications.

From the very beginning of polytechnical training which goes back to the intellectual history of the French Revolution, it has been of a multi-functional, additive nature. Today the polyvalent education, however must be seen as developing from a central point or a focusing point. I would like to explain this by an example. It relates to the multitude of meanings of the term and the skill of 'measuring'. There does not exist one single profession in which measuring would not be part of the job: by means of a gauge, a meter, a counter or by scales, to name only the classical instruments by means of sensory testing techniques or with complicated methods provided by analytical chemistry, where the natural science are concerned; in this connection we should not omit the empirical methods of the social sciences which make use of statistics, or the norms and values of jurisprudence and moral philosophy. To sum it up, we can say that the following factors play an important part when measuring is concerned: the human senses/specific logical abliliticks like relating or comparing/control of processing rules/the ability to handle tools and instruments/the ability to judge results in their causal and final context.

In a small private research and consulting company which I founded my colleagues and I work on the assessment of the subject 'measuring' and of other similar subjects (e.g., dealing with classifications, handling of information) in order to find out whether curricular modules can be built, which might be suitable for a polytechnical and polyvalent basic education on the secondary level which at the same time transmits development of mind, qualification and enlightenment.

The same procedure could and even should be used on the university level. But we have not reached this stage yet, although we are busy collecting material.

With the help of the above example you might realize now the effectiveness of a polyvalent education but I still have to show you what we can learn from this example regarding the value of enlightenment.

Well, some effects will present themselves quite automatically: It can be stated for instance, that following relevant ecological campaigns the use of highly toxic pesticides decreases, a phenomenon that can be illustrated by diagrams from which declining growth rates and even an absolute decrease in the application of toxic substances can be read. Every environment minister would proudly present a graph with such a flattening curvature and he would be absolutely convinced that the measuring result is correct. An ecologist on the other hand, would not direct his attention to the changing curve but to the overall quantity of toxic substances which have accumulated, and their consequences would be a challenge to him. A natural historian would point to the fact that the earth has seen great and dramatic changes of climate in the course of its history which leads him to the conclusion that we have no reason to overestimate the hole in the ozone layer. A biochemist, however, would make analysizes of pollutants and would name those factors for which we are responsible and which we therefore can control.

You see what I am getting at: I want to point out the alienation effect, as Brecht the playwriter has called it. It is the irritation which forces us to come to a halt and look around with "new eyes"; and this is what allows us to accept enlightenment. In my opinion education also means irritation. In scientific terms this would mean that considering the results of research we should reflect on the leading interest of the particular research, the methodologies used, their range of effectiveness and the why and how of their representation.

Of course I know that with this conception I step on delicate ground. When Schwann and Schleiden discovered and defined the cell in 1838/39 at the latest, a development was launched which began to concentrate on the research of the microcosm, on the exploring of an ever diminishing small world. Along the lines of this development more and more knowledge has been accumulated on things and subjects which have become smaller and smaller. Above all, this has led to results in medicine and biology which are just as fantastic as they are frightening.

The nuclear and ecological threat to the earth however, the problems of unemployment which for their part are of a technical-economical nature, the problems arising out of an increasing meaninglessness and loss of identity through socio-cultural desolation: all these phenomena make it clear that we badly need more than to accumulate more and more knowledge and skills in even smaller fields. What we have to learn is to cope with indefiniteness and complexity.

Yet we still have to state that whoever deals with indefiniteness and complexity in one's research work meets with distrust and suspicion. One is blamed for using mysticism instead of mathematics, arbitrary interpretation rather than factual search for the truth. Science as a source of reliable factual knowledge is expected to proceed empirically in a descriptive way and thereafter systemize the enormous number of facts. Therefore engineers often are less creative than one imagines. What they want is a well-structured grating for classified thinking that follows given patterns.

In contrast to this enlightenment is obligated to quit this notion of scientism and is located in a field of scientific philosophy and in strategies of cognition which serve to make constellations transparent and thus make optimum decisions possible.

The purpose of enlightenment therefore is - and this is my fourth thesis - to complement an object-orientated method of thinking the tendency of which is to concentrate on the microdomain, to isolate, to merely state and classify. In contrast, enlightenment does not only mean knowledge but is a way of thinking that helps to see things in their context, in their developing processes and in consideration of changing dimensions and changing fundaments of cognition.

I would like to call this way of thinking change- and subject-orientated. It means that it radiates from the different nodal points of a network or that it aims at them. In the literature I used I found a simple and yet highly convincing example. It is taken from dental medicine: The forms of human teeth can be examined, described and grouped by conceptual categories according to their specific forms and functions (incisors, eyeteeth and molars), or they can be grouped under the heading of human dentition. The individual tooth is divided into crown, neck and root, and its internal structure can be determined down to every single substance. The generic term of tooth is thus subdivided into special terms and defined down to the smallest detail. The method of thinking which is applied here is of an object-orientated, determining nature. It disregards and excludes that the term of 'tooth' comprises more than that: It may not only be understood as incisor, eyetooth and molar, but also as dentition, or as an organ of the digestive system or as epidermatic organ of the skeleton and many other things. You can see that the term 'tooth' can be the starting point for a variety of coherent notions which in their significance go far beyond additive thinking and spread in all directions. To bundle all these meanings it is necessary to constantly think of the permanent correlation of the whole and its parts.

It depends on our intellectual curiosity how we approach the building of a scientific vision of the world, which attitude we take and at which conclusions and perspectives we arrive through studying and relating things to one another.

At this point the crammed curricula of modern study courses will certainly have come to your minds and you are asking me and yourselves how enlightenment can be achieved along the lines of a subject-orientated way of thinking which I just outlined, at what time it should be achieved and how much time it would take.

I do not have any formula for it. One way would be to create polyvalent modules like measuring at the university level. One third of the study course could perhaps consist of modules of this kind. Another way could be the study of for instance Goethe's papers on natural sciences because they show holistic thinking in an exemplary way. Basic studies of ecology could be offered as a further part of the curriculum and one also might consider art exercises in order to train the senses and the awareness of multiple meanings.

Although these study subjects cannot be taken for granted in modern management courses yet, they are no longer rare occurrences. Our way of acting and our civilization can only change to the better if also our way of thinking changes. Managers and business people already have begun to learn that. The relevant methods of learning, valid for technicians and engineers are in principle not different from those for economists. In my opinion, however, considering the great importance of technology in our civilization essential new impulses towards enlightenment could only be given by the technical sciences and the technical professions, if they succeeded in accepting a new approach towards technical training and in helping to make it popular.

Didactical Structure of the Technological Culture

Vanya Georgieva
South West University 'Neofit Rilski'
Engineering Pedagogical Faculty
Alexi Velichkov Street 66, 2700 Blagoevgrad, Bulgaria

Abstract. The technological culture can be and should be viewed on as a relatively independent part of the human culture. Its didactical contents and structure include elements of the technical, organizational, economic, aesthetic, environmental, moral, legal, health and motive cultures. All these can be also viewed on as relatively independent modules.

The technological culture is not only a total sum of these modules. It is a complex dynamic system of mutually interweaving and coinciding elements. This characteristic of the contents of the technological culture arouses many problems to the pedagogues when they make the curriculum in the comprehensive schools.

Keywords. Didactical structure, interweaving and coinciding elements, technological culture

In my country there are three concepts for the forming of technological culture in school. According to the first concept the technological culture is a part of the content of most of the school subjects. According to the second concept the technological culture should be taught as a separate school subject. The third concept is a combination of the other two. Educators who accept this concept think that the technological knowledge that has fundamental and theoretical character should be mastered through the basic school subjects: computing, physics, chemistry, biology, etc.; and the knowledge and skills with practical orientation must be concentrated in one or few school subject named 'Technics and Technology' or 'Technologies'. All the three concepts have their advantages and disadvantages.

In the didactical structure and content of the technological culture components of the technical, organizational, mathematical, economical, aesthetic, environmental, moral, law, health and motion cultures are included. They can be considered as relatively independent modules of the curriculum. This point of view is favouring the educators who say that the technological culture is a part of the content of most of the school subjects that are in the curriculum of the schools for general education. But in lots of cases educators cannot integrate this in the content and it remains simply as a mechanically added appendix to the rest of the teaching material or is soon dropped out.

The technological culture is not only a total sum of the modules listed above. It is a complex system of mutually interweaving and coinciding element. The technological culture can be and should be considered as a relatively independent element of the human culture. This point of view favours the educators who think that it is necessary to have one basic school subject through which the technological basis of modern society to be taught. This school subject will not interfere with the status of the other school subjects and through it a logically grounded unified system will be created.

It will create conditions for writing new textbooks and making new school appliances which will correspond to the specific features of the subject. Except that, it will make necessary a new type of teachers to be trained - the so called 'engineers-pedagogues'. This concept has several disadvantages:

1. It requires a lot of capital in the beginning of the introducing of the new subject.
2. There is a lack of well developed methods of teaching.
3. The teachers who at the present moment teach various production activities that are part of the traditional labour education cannot teach a school subject with a polyvalent teaching content.

At the present moment the majority of votes is given to the third concept which unites all the advantages and some of the disadvantages of the other two. We develop the idea that the didactical structure of technological culture consists of:

– a system of fundamental knowledge of the contemporary scientific technology basis and of the general trends in the development of the main technologies in the different branches of economy.
– a system of approaches, goals and strategies which are part of the modern labour style and the material production.
– a system of general and specific professional labour skills necessary for the development and realization of contemporary technologies.
– a system of norms in the attitude toward technology as a value and comparing it to the other values of society.

The technological culture of modern society cannot be developed through only one school subject and in only one educational degree. It is an object of the pedagogical activity since the earliest childhood. The components of the technological culture mentioned above must be connected with the specific functions of the different school subjects and must be taken into consideration when building and changing the curriculum structure and contents.

But it is also necessary to have one specific school subject that will secure a gradual introducing of the child to the world of technology since the earliest educational age. This school subject should not be made of practical knowledge and skills that are apart of the system of sciences. In this subject the common characteristics of the contemporary technologies must be concentrated and localized. This school subject must have a common name at all levels of education in school and pre-school.

The development of this school subject starts with the answer to the question: Which are the basic ideas from the sphere of technics and technologies that should be included in the content of the subject?

These ideas must be a relatively small number. They must be mutually related and must form the foundations of the contemporary technological culture. We separate the process of forming of the technological culture in school and pre-school into three levels. At every one of these levels several ideas can be taken as basic. They are contained in every concrete theme of the educational material and influence the choice of strategies, methods and techniques of teaching.

In the educational work with children at the age of 5 to 6 as basic ideas are taken:

– the differences between game and labour
– the technical means are an extension of man's arms
– there is a mutual relation between the functions and the construction of the technical means
– the technical means have the following characteristics: polyfunctionality, reliability, economy, ergonomy and aesthetic outlook.

In the education of students from 7 to 11 years the basic ideas are:

– in modern style of labour it is necessary to apply permanently thought and precision
– the computer, the engine and the robot as modern technical means
– contemporary technologies must be with small quantity of operations, must require little energy and be wasteless
– there is a decentralization in the organization of the production process and when solving all kinds of business problems.

In the education of students at the age of 11 to 16 the basic ideas are that:

– there is a technological division of human labour but also a lot of common features in contemporary technologies
– there are module elements (machine elements, mechanics, organizational forms, concrete techniques, etc.) that build the technical objects, the technology and organization of labour
– there is a connection between the professional proficiency and the personal gifts and features of character
– the market relations have an influence on the contemporary technologies
– modern technologies should protect environment.

As it is with the other school subjects the technological culture should be mastered by the students through the learning of facts, images, concepts, terms, regularities, principles, approaches, ideas, rules and norms.

This mastering is done most effectively through a system of activities in which the norms of behaviour in modern business and the methods of research and practical activity in the sphere of different technologies are set. The constructing of a system of activities for understanding the basic ideas of the contemporary technological culture is a problem that the pedagogical technologies have to solve.

A Teaching Strategy to Promote Student Confidence and Creativity in the Design and Prototyping of Digital Electronic Systems

Clive D. Mockford
Loughborough University
Department of Design and Technology
Loughborough LE11 3TU, United Kingdom

Abstract. This paper reports the structure, content and observed success of an innovative teaching strategy in the subject area of digital electronics that has been developed for undergraduate students of Industrial Design and Technology at Loughborough University of Technology.
   The rapid and continuing expansion of the range of devices and tools available to digital electronic systems designers, especially in the area of programmable and microprocessor systems coupled with computer simulation, has prompted this change in teaching strategy. The approach adopted suggests that not all traditional aspects of digital electronics courses need to be included in an introductory syllabus, but that if a focused, limited range of knowledge and skills can be taught in depth then this may enable students to deal more effectively with open ended design situations. The main focus of the teaching programme is a series of lectures and laboratory sessions dealing with the organization, characteristics and applications of the UV erasable EPROM (Electrically Programmable Read Only Memory). The teaching programme seeks to engage students in a systems approach to design, whilst providing a narrow but deep spline of knowledge to facilitate confident design decision making. In this way, student confidence is promoted during the module. Later in the course, students are provided with an opportunity to practice creative technological design through engagement in the design and development of solutions to client based industrial projects. The extent to which the taught knowledge base has been utilized and a systems design strategy adopted in this later project work has been used as a tangible indicator of the success of the teaching programme.

Keywords. Design, digital electronics, innovative teaching strategy, teaching programme, technology, technological knowledge

1 Introduction

The pace of technological change is considerable, particularly in the area of microelectronics, where the use of VLSI (Very Large Scale Integration) technologies has led to the production of a wide range of powerful devices where the unit cost is low and functionality high. Teaching young technological designers in this culture of change makes considerable demands on the teacher. In particular, care is required when designing syllabuses to ensure that students are provided with the appropriate range and depth of knowledge and skills. It would be impossible to incorporate all elements of digital electronic systems into a course, even for a degree course in electronic engineering. In this instance, where the course is in Industrial Design and Technology, a balance between technological knowledge, practical skills and design experiences is required. An inappropriate balance can result in the course providing a range of technological learning experiences that provide students with awareness rather than design capability (Myerson, 1992).

In current professional practice, when developing electronic solutions to design problems, the technological designer is increasingly becoming an assembler of systems and sub-systems, moving only for short periods of working into the component level of activity (Martin et al., 1993). Understanding and using the language of technology, communicating essential details of performance, interconnectivity and interface requirements of systems and sub-systems are vital skills at the higher levels of activity. Yet an education programme that encourages a purely systems approach can lead students towards the adoption of shallow approaches to learning, plugging sub-systems together without understanding the need for a total design concept or without due consideration of signal and electrical compatibility. In this context many students may rely on the acquisition of surface knowledge rather than deep knowledge. As a result, they frequently place themselves in a situation where they do not have sufficient depth of technical knowledge to exercise valid design judgements.

Within the context of an industrial design course, framing a syllabus to ensure that technologically literate and practically competent technological designers are encouraged to develop in a creative environment presents a number of key difficulties:

– what technological knowledge should be taught within the time and resource constraints of the course module, and at what level?
– at what level should the technological knowledge be taught?
– what strategies should be adopted for teaching technological knowledge and associated design skills?
– how to provide and manage opportunities where students can practize creative and successful designing in a technological context?

Engagement in design and technology can be a powerful learning experience in which active learning plays a major role. Active learning suggests students engaging more with their learning rather than being passive, generating specific outcomes. A substantial compromise is required in terms of the knowledge base that is taught if design based, practical activities are to be included. Text books dealing with digital electronics emphasise the largely traditional approach: a substantial knowledge base coupled with a variety of analysis and minimisation techniques for effective circuit design. Many of these approaches work on the principle of minimal circuit design, where redundant elements are eliminated. The advent of VLSI has ensured that circuits can now be designed in a different way, using a non-conventional approach. An example is the problem of implementing a logic expression, stated in mathematical form, as a digital electronic logic circuit. This would previously have been designed using a hierarchical, linear strategy to reduce the mathematical expression to its minimal form, which would then be implemented using a specific, homogeneous logic function to produce the working circuit. As a result of the development of VLSI devices, this logic expression could be implemented in one design step, using an EPROM. An element of redundancy would be evident when using this device, but cost, component size and number identify this implementation as highly appropriate.

Digital electronics is, as a result of technological advances, making use of devices that offer flexibility through programming. Examples are programmable logic arrays (PLA), peripheral interface controllers (PIC) and a wide range of microprocessors. Here, a low cost device, offering high flexibility through programming, can be used to solve often simple, trivial problems which in the past would have required a much more comprehensive approach to design and resulted in a considerably higher circuit board chip count and layout complexity. The idea of implementing a logic design that has not been reduced to its minimal mathematical form or of using a very powerful programmable device to perform a simple task can imply poor design. However, the cost, availability and speed of these devices are factors that can be used to justify the acceptance of the redundancy evident in this approach.

If a conventional approach to teaching is maintained, following the structures adopted by the majority of digital electronic text books then overload is soon reached; the volume of material to be covered is too large and the opportunities when students practice using and applying the knowledge too few in an overcrowded syllabus. Alternative approaches to teaching digital electronics through CAL (Computer Assisted Learning) using simulations of circuits and systems can offer advantages in terms of the amount of time needed by students to complete a scheme of work (Dobson et al., 1994). One consequence of these approaches can be a reduction or complete cessation of practical work. In these circumstances the potential benefits of active learning can be lost. Whilst it is acknowledged that students need to experience and use ECAD (Electronics Computer Aided Design) to reflect modern design practice, a deeper learning opportunity can be lost if a balanced approach is not adopted.

The course that has been developed made a considerable compromise on content, concentrating on high level, detail knowledge in a number of very narrow areas.

Acknowledgement of the acceptability of using technologically sophisticated devices to solve essentially simple problems was a central feature of the teaching. Techniques which have become redundant were largely eliminated from the taught course, replaced by a narrower but more applications-focused set of learning experiences. The programme culminated in the design and prototyping of a system which used an EPROM to solve a complex logic problem. Whilst a systems view was considered important, a focus on the electrical characteristics of input and output connections to logic devices was a key element of the programme, providing a narrow but deep set of knowledge.

The investigation and application of this knowledge was encouraged through practical laboratory work and a number of short, practical design exercises. Learning about individual microelectronic devices through their input and output characteristics was pursued at a systems level. Whilst this typical systems approach was adopted, students were also presented with a conceptual model of the internal organization of selected devices.

2 The Teaching Programme

Students taking this module of the course had completed a foundation course in electronics, containing elements of analogue and digital electronics. The structure and content of the teaching programme that is the subject of this paper are illustrated in figure 1.

Week	Content synopsis (3 hours of student-tutor contact per week)
1, 2	Combinational Logic System Basic gates, boolean algebra, De Morgan's theorem, minimisation techniques using algebraic and Karnuagh mapping methods
3, 4	Sequential Logic Systems Information coding systems JK Flip flops, asynchronos, and synchronos counters Synchronos counter design from first principles
5	Sequential Logic Systems TTL and CMOS counters: range, operation and selection Digital comparitors
6	Memory Devices Static RAM and UV erasable EPROM: organization, operation, selection

7	Memory Devices
	Memory devices to implement a combinational circuit
	Sequential control using memory devices
8, 9, 10	UV EPROM
	Design Project: compass display decoder and driving system

Fig. 1 The structure and content of the course

The programme concentrated on two traditional aspects of digital electronics courses, namely combinational circuits and sequential circuits. The main focus of the programme was a series of activities based around the 2764 UV erasable EPROM device, bringing together the work completed on combinational and sequential circuits.

Using the EPROM as the focus for the course offered a number of potential learning benefits:

- it can be used to implement a conventional combinational circuit without the need for minimisation: one EPROM device can implement a combinational circuit that would otherwise be a multi-chip circuit;
- when used in conjunction with a sequential device such as a binary counter, it can be used as a simple sequential controller;
- address line inputs required appropriate reference to the manufacturers recommended signal conditions, electrical loading etc., encouraging a well considered design approach;
- data output lines required the manufacturers' electrical loading specification to be adhered to, leading to a detailed analysis of loading and associated output stage design;
- timing and chip control signals involved students in wider aspects of digital systems.

A wide range of design problems could be solved using this device as the central building block for the system. Concentrating student effort on using this common element, the EPROM, as the heart of a system ensured that whilst they were tackling an open-ended problem that could be solved in a variety of ways, the means to obtaining a solution was both a familiar device of which they had technical understanding and mastery, and one where they could design and implement a potential solution in a number of different ways.

In addition, the group were all using the same device for the tutor-specified projects towards the end of the course, allowing shared knowledge and understanding to permeate the class. In many cases this led to a deeper understanding of the function of the EPROM and its flexibility. Students also gained a more accurate knowledge of the electrical characteristics of a particular logic family, in this case CMOS, to which peripheral input and output devices must adhere. This information was largely gained from manufacturers' data sheets, providing students with an opportunity to gain confidence in extracting and using this technical information in a design context.

Further, the need to stay within manufacturers' performance specifications was highlighted. Systems which failed to work at the prototype stage were often characterized by a sound design concept but poor electrical implementation. Using the manufacturers' data sheets, relating performance characteristics and configuration to design and development stages provided students with experience of the typical mode of working adopted by a professional engineer.

At the prototyping stage the use of this device allowed students to experience using software tools for programming data into the device. For many students, the need to specify the internal data set for the device prior to prototyping ensured that the circuit design concept had been thoroughly specified, leading to the likelihood that the circuit would require fewer modifications following testing. In this way a more rigorous approach to design was promoted.

3 The Use of the EPROM Within Client Based Design Work

Following this module in digital electronics a number of students undertook major design projects for industrial clients. The success of using this device as the focus for the preceding electronics module was reflected in the number of projects that incorporated it as an appropriate component in the design. Four projects used the EPROM as the heart of their digital system. These projects covered a wide range: an auxiliary lighting control system; a car windscreen LCD sun blind control system; a sports fitness timing system; a timing and control system for conference speakers.

From the author's experience of student-tutor interaction at the design stage of this project work it appeared that students had a firm grasp of the design opportunities that this device offered, both in terms of combinational and sequential circuits. It allowed them to solve electronic design problems that would have otherwise required either a more complex circuit built from discrete components or a less complex circuit constructed using a microcontroller or microprocessor that required complex software to be developed. Also, the time taken to programme the device with an appropriate data set, build the prototype system and test it proved shorter by comparison with other approaches using discrete components. Confidence was also quite high; the students believed it was a simple device they were using and understood in detail the input and output characteristics along with the appropriate control signals that are necessary for its effective use.

This aspect of perceived confidence is worthy of further discussion. Students frequently believe that open ended design problems set by clients require complex solutions. During project work of this nature a clear analysis of what is required is a key factor leading to a successful design outcome. During this project, if students saw the solution to the design problem as a combinational circuit, or a simple control sequence then they had at their disposal enough knowledge and experience concerning the EPROM to actively develop a practical solution. Design work could then proceed. Whilst the knowledge base was very narrow indeed, the use of the EPROM allowed complex functions to be implemented with this device at the heart of the digital system.

From the authors' direct observation of student design work, an element of creativity was evident amongst the range of solutions considered. It was not simply a case of accepting the first potential solution; other strategies for using the EPROM were considered, in many cases leading to the development of a more refined, efficient and more elegant solution.

4 Summary

Through using the EPROM as the focus for learning during the later stages of the teaching programme the level of student confidence when formulating potential solutions to client based design problems appears to have improved. The technical content of the course is now not so extensive as it has been in the past. However, thoroughness has been maintained by considering a number of key concepts in the teaching programme and reinforcing these through the use of the EPROM within practical project work

The overall strategy has been one of establishing a narrow area of electronics knowledge, being taught in depth, and using a very adaptable device as the focus of the later stages of the course. Essential characteristics of digital systems design have thus been maintained.

Students have adopted a broader systems approach to design, specifying the overall requirements for a potential solution prior to implementation in an appropriate technology. In some cases, microprocessor systems would previously have been chosen as the appropriate technology, but following this programme much simpler systems have been adopted by students. It is acknowledged that the dividing line between when to use a microprocessor system in preference to a hard wired system is difficult to identify, for there is clearly a level of complexity in electronics where a microprocessor offers considerable advantages.

In summary two points are particularly worth of note. First, the programme has been well received by students of industrial design and technology who demonstrated confidence within design situations, albeit linked to a narrow technological experience. This has led to the use of the EPROM within project work as a flexible tool, applied creatively. Second, the programme has shown to the tutors that teaching electronics from a narrower knowledge base, linked to a systems perspective can encourage students to adopt a more rigorous approach to learning. It can encourage students to adopt this style of narrow but deep learning in other technological situations. The difficulty with this approach is in specifying the knowledge content of the course and linking this effectively with laboratory and practical work to ensure that students are able to move confidently into more open-ended situations.

References

1. Dobson, E. L., Hill, M.,Turner, J. D.: An evaluation of student response to electronics teaching using a CAL package. In: Eames, I. W., Johnson, A. R. (eds.) Conference proceedings: computer aided learning in engineering. University of Sheffield, 1994
2. Martin, D. J., Coleman, J. M. B.: Developing information skills and system thinking: a resource-based approach to electronic control systems. In: Smith, J. S. (ed.) IDATER93 International Conference on Design and Technology Education Research and Curriculum Development. Loughborough University of Technology, 1993
3. Myerson, J.: Technological change and industrial design education: a review of changing practices on industrial and product design courses in the UK higher education. CNAA, London, 1992

The Ontdekplek

Diny Flierman and Harry Valkenier
Highschool Alkmaar
'Stichting Ontdekplek'
Zoetestraat 11, 2011PP Haarlem, The Netherlands

Abstract. The Ontdekplek (discovery place) first step towards Technology Education. Originally designed to address deprived children; Ontdekplek, now a place for every child. Playful education to become acquainted with basics of technology.

Keywords. Discovery place, technology education, after school programme

1 General Information

The Ontdekplek (discovery place) is a cooperation between primary education and social cultural work. It started as a small programme, but in 1993 it became an independent foundation.
   We call the Ontdekplek the place in Haarlem, where children between 4 and 12 years old can get in touch with the basics of modern technology.
   Due to all kinds of exciting experiments with safe and simple materials the children get acquainted in a very playful manner, with the basics of modern technology, which will be further explained later in their school career and which are used in different trade branches.
   Possible activities are bricklaying with miniature bricks, experiments with light, colour and sound. Building and testing airplanes, helicopters, flying saucers and parachutes. Safely connnecting electricity, burn lights and run machines. Making constructions with paper, building the highest tower, the strongest bridge. Building dams and dikes in streaming water. Photography, working in the darkroom. Last spring we started with the Ontdekplek 10+. It is a special programme for children 10 years and up to build bridges, towers of meccanotype of constructions from raw material. In order to build this they create a miniature production process. Originally the Ontdekplek was developed as a project for deprived children in order to enhance their development and chances for their future education. Now we even reach retarded children as well as very gifted ones.

2 The Ontdekplek in and after School

The Ontdekplek offers a programme in school as well as after school. Some skills can be easier taught in school and applied in the Ontdekplek or some skills can be easier learned in the Ontdekplek and used in school. Also the Ontdekplek offers children, without technology lessons at school, a chance to get aquainted with technology.

In the Ontdekplek (this is the after school programme) children can choose from a long list of activities the activity they want to get involved in. From a central desk we hand out the materials needed as well as simple workbooks. In these workbooks hardly any language is used. They are composed with drawn instructions (we produce these workbooks ourselves). This enables children who cannot read Dutch so well (or not at all either because of learning difficulties or due to their Maroccan or Turkish background) to work on a similar level as other children in their age group. The children work very independently, but when needed further guidance is available.

3 The Mobile Ontdekplek

After a year it became obvious that many children were able to find the Ontdekplek easily, but that we did not reach the deprived children we were initially intended for. They generally remain in their neighbourhoods.

In answer to this we developed the Mobile Ontdekplek. A travelling Ontdekplek, in the sense that materials needed for the Ontdekplek activities would get transported to a school or a community centre, where a temporary Ontdekplek would arise on the spot.

We have preference for a cooperation between primary education and social cultural work since knowledge is only really effective when used in leisure and when it can get expanded on.

The social cultural work offers great opportunities for this. We offer special possibilities for the school as well as for the community centre catering to their specific needs.

Our offer to a school consists of two weeks of technology lessons either with an inclass instruction for one teacher and one group or for three teachers and three groups after which the teachers are able to teach the simple basics of technology. Also the whole team can enroll themselves in an eight-hour-course especially designed for their school in order to enlarge the possible subjects for technology lessons in their school. Material and workbooks are included. We train the social cultural workers and their volunteers. After which they are able to offer at least a six-week-programme. We advise about the content of the programme offered and donate the material needed as well as workbooks for the children. We strongly stimulate the cooperation of the school and the community centre in the same neighbourhood and to attune their activities. The cooperation especially from the social cultural workers whose enthusiasm often motivates the schools to join. The Mobile Ontdekplek has sucessfully ended its two-year-experiment and will develop a storehouse where activity boxes can be lent.

4 Finances

Finances have always been a problem for the Ontdekplek. Since this year we get partially-funded with a grant for small organizations like ours. Some money we make on the sale of the workbooks, is invested in the Ontdekplek.

At this moment we are able to finance 24 professional work hours for the work involving the Ontdekplek and the Mobile Ontdekplek and the development of new activities, which makes it necessary to work also with volunteers with all problems involved.

For the Mobile Ontdekplek we have a budget of Hfl. 8.000,00 to cover costs of material, workbooks and transportation. The schools and the community centres pay a small fee for the use of materials.

At the moment the Ontdekplek itself has no budget for materials, we work with the financial contribution of the children which amounts to Hfl. 3,00 per child per afternoon. We try to keep the financial contribution of the children as low as possible so money will not be an obstacle.

With the little money and professional workhours we have, we reach at least 2.500 children in the Ontdekplek per year and another 1.200 in the first year of the Mobile Ontdekplek.

5 Future Plans

This year three permanent branches of the Ontdekplek have opened in community centres in the northern part of the country.

We received a government grant of Hfl. 50.000,00 to create a Mobile Ontdekplek for the north-west part of the country. A network of interested educators for primary education already exists due to the past graduate course at the Hogeschool Alkmaar, which Harry teaches.

The Mobile Ontdekplek will always try to stimulate the cooperation between primary education and social cultural work. The workbooks, produced by the Ontdekplek, have also been tested in Poland as well as in England and have proved to work well.

We are also working on an efficient English teachers' guide to go along side the workbooks. Naturally the workbooks could be used everywhere in the world since language is not used for instruction.

The Teaching Laboratory as a Meaningful Environment

Matzi Eliahu
Center for Technology Education affiliated with Tel-Aviv University
Department of Education
53 Golomb Street, Holon 58102, Israel

Abstract. The Center for Technology Education Holon (CTEH) has established in 1994 a new and innovated Teaching Laboratory for the Electrical and Electronic (TLEE) subjects. The main reason which led to this specific TLEE is our philosophy that we have to conduct, both theoretical and experimental lessons, in the same environment, whenever we are looking for meaningful teaching, of pre-service students, the following methodology subjects: Electrical; Electronics; Control; Robotics. In order to achieve this goal we equipped the TLEE with: 486DX computers (connected via the Novell network); video camera; VTR, TV, Instructional Learning Units (in: Computer Vision; Computerized Process Control; Computer Analysis and Simulation; Robotics; Microprocessor); Measuring instruments (Scope; DVM, Function Generator; Power Supply; Matrix Board); A powerful LCD overhead display; Multimedia accessories; Internet connection; High speed modem.

According to the mentioned philosophy, the methodology lecturer (with about 30 students), the students while giving their 'lessons' and their instructional teachers (with 16 students in each of the instructional groups), all of them are teaching and using the mentioned facilities of the TLEE.

Keywords. Teaching Laboratory, technology education, methodology

1 Introduction

Teaching a technology subject such as: Electricity, Electronics and Control at high schools and colleges, involves of three sections: Theoretical, Experimental and Constructional.

In the Theoretical Section the students usually getting theoretical knowledge that the teacher is delivering using the classic methods:

- chalk & board
- transparencies

In the Experimental Section the students are conducting fundamental experiments, usually with commercial didactic units, which should proof the concepts, facts and knowledge that they have learnt at the theoretical section.

In the Constructional Section the students are usually building, from scratch, basic electronic circuits such as: light detector; organ; secret locker; audio amplifier, etc.

2 Facts

In most of the cases the Theoretical Teacher is not the one that conducts the Experimental Section neither the Constructional Section. The Experimental and the Constructional Sections, together, have the same learning hours as the Theoretical one (between four to five weekly hours). Even, both the theoretical and the experimental teachers, will have good cooperation with each other, still the fact that they are not the same person will reduce the efficiency of their mutual teaching. In school reality this cooperation is partial one (some times even without face to face regular meeting, just using phones or notes) and occasionally even not existing at all. One can argue that the main reason for this diversity between the teachers is a technical one (i.e., problems of "putting" teachers in the right lessons and at the right time).

The author found from his experience that, in most of the cases, the actual reason is the difference in teachers' qualifications. Usually the theoretical teacher will have a BSc degree (with or without teaching diploma) while the experimental and constructional teachers will have a Technician degree or a Technical Engineering degree.

3 The Problem

There is a gap between the mental capacity, of students at their 10th-11th grade, and the cognitive demands needed to learn Electricity, Electronics and Control – i.e., their cognitive stage, which is the Concrete Formal one, makes the learning process of formal concepts very difficult. Concepts that are very often needed in these subjects.

4 Target

We have to narrow this students' gap as much as we can. There are different methods to achieve this goal, and the most efficient ones are:

- visualization
- experiencing
- construction.

5 The Solution

So, what can we do about it?!
First, where it is possible, the theoretical teacher should conduct also the experimental lessons.

When it is not possible, both the theoretical and the experimental teacher should cooperate very closely using fixed, face to face meeting, each week. Consequently the theoretical teacher should use visualization elements as an integral part of his lesson. The visualization should be meaningful by involving the students in an Interactive Visualization manner. A lesson that incorporates Interactive Visualization can be more effective and would help in narrowing the mentioned gap. It is obvious that the teacher should be prepared and should have enough competency in using and operating the Interactive Visualization.

It is not simple at all when you have to focus the attention of about thirty students in an instructional unit; measuring instruments; component(s), etc. Hence, the earlier we can start and prepare the teacher to use this kind of teaching method, i.e., the Interactive Visualization, we will make it easier for him later.

6 Timing

The question is when actually to start? Our suggestion is to start at the methodological courses. The methodological courses give the student his first knowledge about how to teach technology subjects such as: Electricity, Electronics and Control. Actually the methodological courses are challenging its teachers with the same problems that exist at the high schools (i.e., the diversity between the theoretical lecturer and the instructional one).

The academic course also composed of three sections: Theoretical, Instructional and Actual Experiencing (class teaching). Usually the Theoretical part is given by a PhD lecturer while the Instructional and Actual Experiencing parts are both given by an Instructional Teacher that usually is a very experienced one with an Engineering degree (BSc) at least. There are logistic problems in the educational departments, at the universities, in Israel and in the western world. These educational departments usually do not have suitable Instructional lab(s) for teaching technological subjects. A fact that makes every desire to use demonstration or visualization an unrealizable one.

7 Implementation

The author joined the Educational department at the Center of Technology Education Holon (CTEH) affiliated with Tel-Aviv University, in 1994, as a Methodology lecturer of Electricity, Electronics and Control subjects.

Throughout the recent academic year (1994), a new and innovated teaching and instructing lab was constructed by the author in the Educational department of the CTEH. This lab solved the mentioned problem. At the current academic year, 1994, the author has experienced in combining the Interactive Visualization, and some times even combining Experimentation, as an integral part of the theoretical section of the Methodology courses. He used to meet 30 students, at the new teaching lab, which contains: computers connected in LAN; instructional units; measuring instrumentation; audio visual instrumentations and a high power LCD projector.

8 Examples

a. Using computerized Packages to Design & Analyse Electrical and Electronic circuits. Such as: Mcap IV and Pspice (for windows).

b. Using Audio Visual Instrumentations for displaying measurement instrumentation during conducting an experiment.

c. Fundamentals in Computer Vision: self-learning at home; explanation of difficult issues by the teacher at the class; team experimentation.

Education to Teamwork in Building Technology
MHPO - Method Holistic Participation – A Contribution to Technology Education

Peter Schmid
Eindhoven University of Technology
Faculty of Architecture and Building Science
P.O. Box 513, 5600 MB Eindhoven, The Netherlands

Abstract. Technology mostly is characterized by a high degree of complexity. This demands in practice the cooperation and collaboration of various experts. Scientists, technologists, but also very often specialists in the fields of finance, economy or commerce, politics, sociality and management together are the participants in solving technological problems. Moreover we may not forget, that various governmental levels and particularly the clients are shaping the final technological processes and results.

In a time of a growing environmental consciousness about the ecological problems, which are also created by the way in which technology is developed and applied, we have to consider a (more) harmonious 'collaboration' with the energies and resources or shortly the natural base on and within which (human) life can find its vital conditions.

Former cultures knew consultation processes in order to solve their problems. Konrad Wachsmann and Walter Gropius introduced in the fourties a method for design and development of buildings and building products, which considers highly the complex 'natur' of this process.

Based on this 'pioneering' approaches the author continued and developed these inputs to the MHP - Method Holistic Participation. This method was proofed in many workshops in several countries and with the most different groups beside the use of it, with students in different ages including also 14- and 18-year old pupils and students. The method was applied in work with experts in several fields, like physics or educators and members of communes, who wanted to build a common neighbourhood and politicians as well.

In this paper it will be shown, how the method is structured and which results - by examples - can be reached. A list with references about the teamworkshops and MHP-sessions within the Eindhoven University as well as in practise around, with which quite a lot of experiences could be gained, and some fruitful examples will conclude this contribution.

Keywords. Method Holistic Participation (MHPO), technology, technology education, design decision support

1 MHP - Method Holistic Participation Research and Experience on Team Work During the Decades

1.1 Remark

Design Decision Support Systems always are closely related to the cooperation of more (than single) people and mostly even of several groups of people. Hence teamwork is an important part in order to support (design) decisions or systematical and methodical collaboration and cooperation or participation. A system(atic) approach towards these participatory processes will be given in this paper.

1.2 Introduction

All human working processes and all results within our culture are more or less based on cooperation. Even the existence of a hermit depends on the fact, that other individuals in society work together. It may be, that cooperation and consensus sometimes or often automatically and instinctively takes place like in the case of simple things and within so-called primitive cultures. In our time and in our industrialized world with a high complexity and many complicated circumstances, nothing is less natural, than automatic cooperation and consensus including design, planning and building activities.

In his video presentation 'The Global Brain' Peter Russel shows the necessity of working together on various levels, if we want to survive. Therefore in the context of design, certainly design participation, it is an actual question to apply efficient methods and techniques to reach the aims of an harmonious working process and integrated results. In this brief contribution to holistic participation we will define the terms we are discussing. In addition to the starting points we have to clarify the aims. At the base of the roots of cooperation, possibilities and needs it is essential to pay attention to the method. Examples of actual workshops give an illustration of the method and the aims. Future applications are of interest for everybody who is participating in the one or other cooperative process. A summarizing conclusion will underline the opportunities we find in the discussed holistic participation, along with many benefits.

1.3 Definitions

Holism is a theory of wholeness, Gansheid, totality, originating in Greece. Holistic is a balanced integrated complexity. Participating from latin 'pars', a part, means to be and to act as a part of a larger order or higher totality or wholeness.

In the context of an holistic participation we find some more terms important to handle the very subject:

Cooperation   working together generally
Teamwork   special structured cooperation

Coordination   multidisciplinary, interdisciplinary organization
                       (Synthezise)
Integration   composition, gestaltung or coordination
Consensus   common (parallel) sense and complementary addition
                       to each other
Method   clear, systematic and efficient way for problemsolving
Process   the path, the way (including causes and conditions)
Result   the goal, the effect of all participating influences

1.4 Aims

According to Zen and even some other philosophies dealing with selfrealization 'The Path and the Goal is One'. Hence it is useful for all participants of a team or a working group to first try and cooperate in a convenient way, to be able to handle conflicts, to work together stimulating, enjoyable and efficient, ... and second to try and reach satisfying solutions, which fit the tasks, problems or questions as well as the needs and intentions of the participants and/or clients etc. The general starting point for holistic participation is the fact that in order to survive, there are innumerable tasks we have to fulfil in our social life together in large and small groups or teams. Design in the framework of shaping our built environment belongs or leads to the conditions responsible for our more or less common existence. So the aims of the holistic participation method are:

- To include at least in principle and/or in a representative way all participants.

- The basis qualities of human and ecological conditions or in other words an integral bio-logical approach carried by the main characteristics of 'bios' and 'logos' belongs to an holistic participation.

1.5 Background

Konrad Wachsmann and Walter Gropius introduced a certain teamwork method for the development of complex building concepts already in the fourties. They indeed never mentioned the source(s) of this system. The author who continued and developed this method in tens of cases, actually found the probably base of the essence of the method in at least the old Red-Indian and Indonesian way to solve problems in a 'democratic' and harmonious way within their tribes. While the current environmental and health problems (Sick Building Syndrome - SBS) get more and more attention this becomes a hot item in our whole society.

Integral, holistic functioning is a natural thing in the process of the nature and in the nature as a wholeness. Balance or a dynamic balance is a result of a holistic or again and again an integration focused process. From the middle ages we know about the mystic seven synchronities ('Die sieben Gleichzeitigkeiten'). It was said, that a human being is able - at least in top condition - to see or remember seven items at the same time in order to combine them creatively to a new thing, like an artistic or scientific product or an architectural and building design. So, an individual person even can exercise this synchronizing method or technique systematically to reach balanced results. Today everyone is dependent on a lot of influences and factors. Due to the rise of many specialities in our divided working processes with so many participants even in case of the design processes (formerly very individual and personal). Even everyone is dependent in a way from everyone. In building techniques - and sciences - for the first time - a horizontal (democratic) structure of the various components responsible for the final built results, was introduced and this introduction can be seen as a turning point like the title of Wachsmann book form 1958, first edited in Germany 'Wendepunkt im Bauen'. In the fifties there was hardly any consciousness on the ecological aspects, the specific human factors and environmental problems in the world of civilization. Therefore we had to add all these new factors in this first total view of building components which was brought by Konrad Wachsmann and supported by Walter Gropius. Finally these additions led to the holistic metamodel of an integral biological architecture.

1.6 The Method

The principle of the holistic participation method works as follows:

1. The (main) task or problem which in case of designs always is a complex one, has to be divided into the various partial problems or aspects - according to the ideas of the participants.
2. The whole team has to be divided into small working groups and it is ideal to have three members in each such a group.
3. After these dividing processes, discussed with the whole team, each small working group goes on to investigate one of the partial problems separately.
4. After this investigation period all the team members come together again to inform each other by chosen speakers, who afterwards will change with others to give to everybody the opportunity to speak in a representative way.
5. After this information, discussion and consultation period all working groups change the subject and continue to investigate.
6. This play or game will continue as long as necessary to give everyone the chance to investigate each of the partial problems. At that point the cycle is finished. It is important to work from a more abstract and general way to a more concrete and specific one. Continuation of the work of others, additional work, the use of mental stealing and the creativity of conflict gets an even important place here.
7. There is always a possibility to structure the participants, the partial problems and the time in such a way, that the just explained integration process can be organized and managed.
8. This 'rotating' or 'weaving' process can be continued even for a second part of a period of making a design ready for realization and even realization itself can be handled in a similar way.

9. Consultants (from outside the team) can at least participate during the discussions as early as possible. Of course it is advisable to use audiovisual methods, models and clear text, sketches, drawings etcetera etcetera for deeper research as well as better information and presentation.
10. In ideal circumstances and with a well trained team the teamwork process is extremely enjoyable and the result in a way will come without effort. Side by side with the 'serious' applications of the method it might be good to apply some socio-psychological and creativity games.

1.7 Actual Work

As already in the International Design Participation Conference at the Eindhoven University of Technology in April 1985 presented, the - nowadays MHP - Method Holistic Participation called - way to collaborate, is more and more in use. The Method Holistic Participation is within the Faculty of Architecture and Building Sciences, especially in the (Sub)Department of Building Construction and Realization used as well as in many workshops in several countries (sometimes the Peter Schmid - Method called) for students and practitioners. Moreover the method was also applied in other fields, like education, physics and even politics.

I had my own training with Konrad Wachsmann in the framework of the International Academic Summer Course in Salzburg in the years 1956, 1957, 1958, 1959 and 1960. After this development we applied the teamwork method many, many times with various subjects and groups of different capability and size and in several countries. In the education programmes of the Eindhoven University of Technology I have used the method many times for 'teamwork and integration' exercises groups of students from abroad or the European Delft Workshops 'The Architecture of an Uncertain Future' and people who want to live together a little bit like in an old village but as a new community, people who look for an ecological and healthy environment, for a bio-logical architecture, for a really human home or shelter or for a more self supporting life, building and living environment know to find my help for a preparation in order to participate and design. Even politicians - confronted with the actual ecological problems such as environmental pollution recently were guided in a workshop on base of the described method. So we see a growing number of interested participants, who like to participate methodically. Most recently one of the sub-departments of our faculty decided to introduce the students to the disciplines of this department in a first project which runs on base of the MHP. More and more the method is also applied in practical fields of mainly research and design.

1.8 Future

The MHP can get a place and meaning in the fields of the most urgent problems and within the discussions concerning SBS and ecological disaster (as significantly pointed out in already so many scientific reports and conferences).

Collaboration and cooperation is needed more than ever before - although we can find examples and models already long ago. Two main problems can be answered by using the method - as already proved in several cases: the facing of the environmental demands for building activities, because of the ecological crisis -worldwide combined with the SBS and the necessity for the different (power) groups to come together in order to reach some consensus for our common survival. The paper presentation will be enriched by many illustrations.

After looking far back to the medieval simultaneity of a creative process, the experience of the Wachsmann-Seminars and the facts of the very recent participation workshops we also may look forward to the future of design participation. It is a matter of fact that finally only a conscious and harmonious participation leads to a good cooperation and only 'a good cooperation' makes us survive. A lot of learning processes bring us to a higher level and quality of existence and make us more conscious about our basic needs and aims. One of the main conditions to reach these mentioned qualities probably will be methodic cooperation. In fact the method could be applied to every thinkable task in the field of building design as well as in all other disciplines. I could imagine, that like in an orchestra various teams on various subjects working together will design and build an harmonious environment which fits in a democratic way the needs of all participants and people. The design for a 'House for Another Future', a prototype and pilot project fully based on a Health and Environment Scenario will be worked out by means of the MHP.

1.9 Concluding Summary

The starting point for a holistic design participation process in architecture is based on a methodically, enjoyable and stimulating cooperation in order to finally reach efficient results. The necessary consensus of the participants may be fed by the creativity of conflict, but all important aspects, components, and factors have to get the chance to influence as relevant causes the effects. During all times mankind had to handle a lot of - sometimes complicated - ingredients to shape a complex totality on various levels. Synchronity, simultaneity, balance, equilibrium was always an aim in the design processes. First Wachsmann introduced a teamwork which fits these demands (Forderungen), by means of a rhythmic working - and discussing process, all research and information can be done and given systematically. A lot of training of this kind have been done and developed in many countries. Hence there is already enough experience from the past and very recent past for further applications in the future. The method can even be expanded to several other fields. Hopefully this method holistic-participation can be a practical contribution for our built environment with a higher responsibility in relation to the human beings and the environment and generally for a more harmonious cooperation.

1.10 Important Remark

Similar contribution focused on the theme are expected by Cees Duijvestein, Joos Hamer, Howard Liddell, Frederica Miller, Ottokar Uhl.

References

1. Schmid, P.: Bio-logische Architektur, (Kapitel über Teamwork) Verlagsgesellschaft Rudolf Müller, Köln-Braunsfeld 1983
2. Schmid, P., Bax, Th., Dinjens, P., Geest, J. van, Mrevlje-Polak, D.: Contribution in Participation - Open Day, 1983 GOM T.H.E. Eindhoven 1982
3. Beekman, P., Schmid, P.: Teamwork Postzegel, AM & AS1, T.H.E., Eindhoven 1983
4.Schmid, P.: Holistic Participation, Open House International - Housing and the Built Environment, Theories, Design Methods and Practise. pp. 10-24 (1985)
5. Schmid, P.: Holistic Participation, Newsletter T.H. Eindhoven, DPC-Design Participation Conference, 12, (1985)
6. Schmid, P.: Holistic Participation (abstract) on DPC-International Design Participation Conference, Eindhoven 1985.
7. Schmid, P.: Holistic Participation. In: M. R. Beheshti (ed.), Proceedings Design Coalition Team - DPC-Design Participation Conference 1,
pp. 272-290, Eindhoven 1985
8. Schmid, P.: Holistic Participation in Design Coalition Team. Vol. One, Vol. Three. Proceedings of the International Design Participation Confernce. Eindhoven University of Technology April 22-24 (1985)
9. Schmid, P.: New Architecture for the Future, Round Table Architects Colonia, Germany Global Cooperation Architecture, Vienna, Austria in Visions of a Better World, A United Nations Peace Messenger Publications, edited by the Brahma Kumaris, World Spiritual University, Fall 1993, pp. 58-59 (1993)

Management and Innovation

Haris Papoutsakis
Technical Education Institute of Heraklion (T.E.I.)
Electrical Engineering Department
Crete, GR-715 00, Greece

Abstract. One of the key priorities of creative and realistic management – a today's must for every enterprise – is the continuous search of the *innovation*. Envisaging the future and analysing the market, either in the traditional way or using modern methods like SWOT Analysis, can be of great assistance. They offer a combined knowledge, regarding the company and the business environment, its departments, the products or services and the competition. They help managers recognize the strong and weak points of their organization, and be better prepared to face the opportunities and threats just ahead of them. Discipline in strategy, diversified for the different types of businesses, is also important. A unique case study, with more negative than positive hints, is quoted.

Keywords. Innovation, market analysis, competition, strengths-weaknesses-opportunities-threats: SWOT analysis, combining strategies, transforming or neutralising strategies, competitive situation analysis, company evaluation

1 Envisage

Too often, profitable companies become comfortable companies and then ... they are no longer profitable.

Every person, who around the year 2010 will be part of the active buying population, has already been born. What is then expected from today's businessmen, is to locate *now* their needs for the year 2010 and make sure they can fulfill most of them. Demographic evolutions are of major importance to the way modern business activities are long-term planned.

Envisage is one of the characteristics corporate leaders (Presidents, Managing Directors or CEO's) are expected to show. But very often in everyday's business practice, other staff members, from the bottom to the top range of the hierarchy, have proved they can also have envisages.

1.1 ATT and the PABX

It was in 1909, when a statistic analyser of the big American telephone company ATT, was studying the statistic diagrams of the American public telephone needs for the last 15 years. Statistic data projection into the near future indicated that by 1920, the company should have to employ *one telephone operator for every American family using a telephone!!*

The message was very clear. ATT did not lose time. Within two years the first PABX (Public Automatic Branch Exchange) was available for the market. What followed, is known to everybody.

1.2 Mail Order

All right answers seem very obvious ... afterwards. As Peter Drucker mentions, everybody believed, around 1900, that a promise for money return, following a purchase by mail, if customer was not fully satisfied by the product, was not in favour of the merchant. The slogan *satisfaction guaranteed or your money back*, was only going to result to the merchant's bankruptcy. This did not stop American businessmen Sears & Roebuck to start mail order sales, which today have reached enormous turn-overs.

1.3 Super Markets and Shopping Centers

It was also known to everybody, around 1925, that people preferred to buy from small specialized shops, spread over the city's market area. Nobody could dare to propose anything like today's *Super Markets*. Also everybody knew, until the late 1950s, that people preferred to buy from the city center. And today we see a very strong trend of both *Super Markets and Shopping Centers* to be located far from the city's center, and still attract the vast majority of the consumers.

1.4 Mass Media

What we today call *Mass Media,* would not have been there, if two major innovations had not taken place at the end of the last century (around 1890). The first was technological: the printing press that allowed newspapers and magazines to be printed at high speeds, superior quality and yet low prices. The second was social-economic: advertisement invasion which started with the 'New York Times' and 'New York World' as well as the Randolph Hearst's media. Advertising, through marketing mechanisms, provided to the media the funds that allowed them to be sold at reasonably low prices.

1.5 Polaroid

Polaroid's success, with the built-in photo developing, is an excellent example of a leader's way of thinking. Built-in developing did not answer an existing customer requirement. People were happy with the standard way of film developing through their photographer. But the time was there for an innovation to come and offer something special to a saturated market. And Polaroid was there to come out victorious.

1.6 Xerox

Very similar to the above is Rank Xeroxes case. Everybody was more than happy with the 2-3 carbon copies (and we still use the cc abbreviation!!) that every good typist could produce and there comes Xerox, with the new copying machine that really changed our life. The innovation, and the created patent have offered Xerox the dominating leading role in the industry that their competitors have really fought very hard to just doubt.

1.7 Robot

The Japanese are today leaders in the robot industry. The reason: they simply studied very carefully the demographic data of the 1960s. It was obvious then, that there was an increasing trend, to families, to offer university education to their children. As a result, the number of workers, who at the 1980s and 1990s would be available for the traditional labour jobs would be dramatically reduced. Everybody was aware of the phenomenon, but it was only the Japanese who took the proper business actions, and here they are leading the industry today.

It is obvious from the above, that an *innovation* must not only be the transaction of a customer need into a *business opportunity*. Far and beyond of that, it has to provide the basis, for these needs to be adjusted to a more general development that occurs and which permits to a number of ideas, rejected up-to-date, to win broad customer acceptance.

2 The Market

Conservative management would prefer the easier and safer way of running a business through well established products and methods, instead of being exposed to the rather insecure and expensive market explorations, offering uncertain results. This could very well be the aspect in many of ... last century enterprises, or even those of the early 1900s. Could this aspect apply into today's world of increasing competition and economic uncertainty?

Modern companies today *have to* look for innovation, not only in order to benefit from the results, but for one more, not so obvious, reason: searching for the innovation, and the fever that it brings with, keeps the entire company into continuous touch with creativity and development. It means a continuous effort to try and benefit from something new.

2.1 Market Analysis

Analysis for the sake of analysis accomplishes little.
Analysis must serve as a basis for action.

Is there a real need for a *market analysis?* What is the difference between market analysis and the well-known to everybody *market research?* The basic reasons for contacting a market analysis is to collect, analyse and finally benefit from a whole piece of information, through which company management will try to search and locate business *opportunities or threats.*

In accordance, these located opportunities and threats, could in one respect be considered as the limits of expected company achievements. And it is within these limits that managers shall have to frame their department objectives. In other words, market analysis first aims into a historic study, concerning the market conditions, and then tries to better evaluate and locate the trends and prospectives expected to appear within the time limits of strategic planning. A market analysis *must* cover - without always being limited to - the following points:

- an analysis of the present situation
- market potential and limits
- competition
- the company strengths & weaknesses
- opportunities and threats

Present Situation Analysis. An analysis of the present situation can be favourable in many ways. It helps company management recognize and better evaluate the most important strong points of the company. It allows all management staff, and not only, express their contradicting views, on a number of proposed innovating ideas. A number of ideas, remaining vague and not fully identified up until now, are better understood. Upon accomplishing the analysis, a complete and realistic description of the company position into its business environment will be in our disposal. We shall have an answer to the question, 'Where do we stand now?'

The analysis, especially when dealing with companies with considerable market shares, has to be extensive so that it covers the social, political, ecological and economical environment of the company. This is what we call *business environment*, as opposed to what is usually referred to as the *market*. The collected data allow management to set the frame into which the company must include its business activities and targets. In another way, minimums and maximums, for company achievements, can be established.

Two historic examples, with notable failure in analysing the present situation should be quoted. The first has to do with *the 1973s oil crisis*. It is hard to explain how unready the western industrial world has been proved in handling the crisis. None (or maybe only one of the seven major oil companies) had realized the increased importance oil drilling obtained against the favoured (during the 1960s) oil distillation and distribution. The second is the well-known case of the *American car industry,* who has been terribly late in finally realizing the dramatic change in customers trends towards smaller size cars. They have been following a wrong, short-term earnings policy, and they almost lost the train to their Japanese competitors.

Market Potential and Limits. In order to better evaluate the *market potential,* described above as a business environment, our analysis should cover:

a. The total market size as well as the company sales increase rate. It is important to know though if sales increases are positive. Because it could very well be the case where a sales increase will not be enough, if let's say, total market is also increasing, but at a higher rate and thus company is losing market share!!
b. *Market segmentation,* based in healthy and customer oriented criteria.
c. A number of market characteristics, like seasonalities, geographic spread and parallel markets where additional sales could be realised.
d. Technological aspect of the market as related to the circle of life of our products and the risks from competitors using advanced technology.
e. Preventive measures against unpleasant market situations, like an economy crisis, a monopoly-type supplier etc.
f. And last but not least, an analysis of previously adapted policies and an evaluation of their results, in an effort to capitalize from our successes, and avoid making the same mistakes.

Competition. It is in every company's strong interest to build, but also to maintain, a dynamic *competitive position* into the business environment. Every company should also protect all those strong points which have allowed certain market shares to its products and significant profits to its share-holders. Therefore the competition analysis should cover:

a. The historic and dynamic presence of all company products into the relevant market segments, including statistic data of their performance in certain time periods, as well as a forecast of their future performance.
b. A comparison of the company position, versus its main competitors. Competitors should be analysed both as a whole, as well as each one individually. Attention is required to information concerning:

– competitor's market shares (analysed per product, customer or market segments etc.)
- forecast on competitor's changes in strategy

– pricing policy
– service offered (very important for industrial durable products).

The *competition analysis* should be done very carefully and following rigid rules. Very often marketing staff tends to overestimate the competition. They have a tendency to recognise much more benefits and strong points, than the ones really existing. They imagine that the competition has better products, better distribution channels, better service and better advertising campaign!! Up to a certain extend this is normal and human. But it is also important to have a clear and fair picture of the competition if we want to benefit from the comparison to follow.

The remaining two points, in order to complete the list set up at the end of paragraph 2.1, in relation to market analysis, have lately earned such a unique importance, that will be seperately analysed in the following paragraph.

3 SWOT Analysis

SWOT (Strengths, Weaknesses, Opportunities, Threats) Analysis is an evaluation method of business alertness. In order to draw an effective strategic plan, managers have to adjust their departments objectives, to the available company resources. This could only happen, when the strong and weak points of the company are clearly identified. Recognizing and evaluating the strengths and weaknesses as well as the opportunities and threats, during business planning process, is called SWOT Analysis, and is illustrated in the following table.

Table 1 SWOT Analysis Matrix

	PRESENT SITUATION	FUTURE SITUATION
GOOD	STRENGTHS	OPPORTUNITIES
BAD	WEAKNESSES	THREATS

According to the application level, SWOT Analysis can be used, by experienced staff-members or managers, for either:

1. Analyse the competitive situation of a company department, a product or a service
2. Evaluate the overall position of a company or an organization. In both cases the search for *innovation* is still the primary objective.
3. SWOT Analysis of the Competitive Situation of a Company Department, a Product or a Service

At the course of strategic planning no matter whether it has to do with a company department, a product or a service, strategic options must be considered. These options can better be recognized by evaluating and controlling both the interior and the exterior environment of the company.

It is at this particular area, that SWOT Analysis can be of significant importance. It helps answering questions like, 'Where do we stand now?' and 'Where do we want to be?' At the course of this analysis, special emphasis should be given at the extend that the company is both answering to customer needs, and main competitors.

Due to the simplicity of the method, it is easy for unexperienced staff-members to conclude a very vague or general SWOT Analysis, contributing very little to the initial requirement. To avoid this certain rules have to be considered:

a. SWOT Analysis accomplishes very little when applied to a problem as a whole. It is much more effective when *focused* to certain areas, such as certain market segment, certain customer group, or certain competitor.
b. SWOT Analysis better pays off when applied by a *team* of the company staff. The information collected by a team is always more accurate than anything the most experienced individual could contribute to.
c. Our analysis has always to be *customer oriented.* Especially when we analyse the strengths and weaknesses. Any strength or weakness that our customers do not recognize as such, or do not considerably value, should not be included in our analysis. This way we are obliged to think in terms of our customers, which of course is the number one rule of marketers. For example, the vague statement, "We are a big company" proposed as a strength, could be further analysed to the following real strengths: "We offer a great product mix", "Our size, is a guarantee for our customers", but could also result to the following, very important, weaknesses: "Bureaucracy", "Luck of communication, at a personal level, with our customers"
d. Opportunities and threats are only real when they are totally *independent of the company environment.* That means that they have to be there, no matter if our company exists or not. There is always the risk for certain staff-members to put into the opportunities 'box' their ideas for certain company policies, and thus later to manifest as prophets!! 'Price reduction' is a policy, not an opportunity. Opportunity, in this particular case, is the existence of an unexploited market segment, which is price sensitive.

With the above rules in mind, SWOT Analysis does allow us to develop two specific strategies as demonstrated in table 2.
Combining Strategies: The emphasis here is to combine the allocated strengths, with the opportunities, offered by the environment. But one shall always be aware of the danger to have a weakness combined with a threat!!! This is an extreme, severe situation that one must be prepared to face.

Table 2 A customer oriented, SWOT Analysis Matrix

STRENGTHS * ← WEAKNESSES *	
Combining Strategies ← Transforming Strategies	
OPPORTUNITIES **	THREATS **
	Transforming or Neutralizing Strategies

   * Must be recognized by the customer
  ** Must be independent of the company environment

Transforming or Neutralizing Strategies: Attention here is required upon examining the weaknesses and threats, in an effort to create strategies that could either neutralize them, or even better, transform the weaknesses into strengths and the threats into opportunities. Needless to mention here, how more difficult developing a transforming strategy is!!!

The following table gives an example of a SWOT Analysis of the competitive situation of a company department, a product or a service.

Table 3 SWOT Analysis of the competitive situation

	PRESENT SITUATION	FUTURE SITUATION
	STRENGTHS	OPPORTUNITIES
GOOD	* 24 hour service * Customer tailored services * Easy approach to the company (parking facilities)	* Unexploited Market segments (i.e., price-sensitive) * New tax-relief law for existing product
	WEAKNESSES	THREATS
BAD	* Not registered TRADE Mark (Not even well known among the consumers)	* Possible future changes in Mktg terms with suppliers * Increase of direct competitors * Possible economic crisis

3.2. SWOT Analysis for Company Evaluation

In case we are interested in evaluating a company as a whole, SWOT Analysis can be a valuable tool. But we should follow a different approach. The main differences from the method described in paragraph above are:

a. The analysis involves the *interior* of the company only. Not the business environment, not even to weaknesses and threats related to the customers.
b. The analysis is applied at *company level* and not to a department or a product or service.

Despite the differences the *team* approach is also very essential here. An example of SWOT Analysis for the evaluation of a company is given in the table following:

Table 4 SWOT Analysis for company evaluation

	PRESENT SITUATION	FUTURE SITUATION
	STRENGTHS	OPPORTUNITIES
GOOD	* Capable staff & professionals	* Creation of new TRADE Mktg Dpt
	* Luck of bureaucracy	* Unexploited solid assets
	WEAKNESSES	THREATS
	* Luck of driving force	* Drastic reduction of R&D expenses
BAD	* Limited network of suppliers	* Product into a declining phase
	* Authoritative Management	

In Table 5 of next page, one can find a list of subjects which should need further investigation, while applying SWOT Analysis for the sake of company evaluation. It is obvious, that the list is long, but it only indicates subjects one might have to consider. Upon closing the subject here, it is useful to be pointed out that *SWOT Analysis by itself is not enough for a company evaluation.* It should only be considered as one of the tools to be used for this purpose. An important tool, providing the selection for further analysis and development.

Table 5 Subjects to be discussed during company investigation

AREA	SUBJECT
1. MANAGEMENT	a. Organogramme b. Management style and decision taking procedure c. Extend of standardization procedures d. Programming and Control Systems
2. PERSONNEL	a. Employees attitude b. Technical skills c. Administrative capabilities of managers d. Training
3. MARKETING	a. Participation in Marketing planning b. Consumer needs knowledge c. The size of the offered product line d. Product or service quality e. Company image f. Market segmentation
4. TECHNICAL DPT	a. Funds for technical skills b. Training and new hires c. Research & Development d. Information on new technology
5. FINANCING	a. Company economic situation b. Profits per market segments c. Extend of development

4 Discipline in Strategy

It has been emphasized that seeking for the *innovation* means a continuous effort to try and benefit from everything new in our business area. Of course, this more or less 'natural' trend must be controlled, in the frame of company strategies, which – in their turn– are determined by its market position, and the general market conditions.

It is obvious that *multinational corporations* with very big market shares and great product mix pay the appropriate attention to the need of *innovation*. Companies in the area of services, and those in the area of marketing durable or fast moving consumer products, tend to have a different approach towards *innovation*.

Companies with very strong leading role in their market, i.e., like IBM or SONY, are not always very keen to apply new technologies in certain products. It is obviously because they do not want to disturb sales of already successful products, at that particular moment. They usually wait for their previous investments to fully pay off, before they kill them, themselves. On the other hand they are very much interested in every innovation leading to cost reduction, since this is the best way to keep new, small competitors out of the market.

Above described policy, by no way means that *multinational corporations* are not always eager to respond and even push for innovations, when forced by the competition or drastic changes in consumer trends.

On the other hand companies in the area of *fast moving consumer products* are those which are in a continuous seek of new production methods as well as marketing and distribution ones. This is exactly the area where *marketing innovations* aim to always better ways of answering – with imagination and efficiency – to the multiple changes taking place into the social and economic environments of a company as well as into its customers buying habits. Which habits are marketing and innovation driven!!! The McNair theory for the 'Wheel of Retailing' in full action!!!

The case of *companies offering services* only – or sometimes products and services – is quite different. Service is an abstract concept with many invisible elements which require difficult and expensive market research or even analysis. This is why there is no innovation barrage in their market. Instead of innovations we notice many ... 'imitations' in a strong, competitive environment.

Everybody realizes how saturated the airlines market is. They are all offering the same services under diligently different names in their efforts to offer something more to their passengers. The same situation is noticed to the banking or insurance companies area, where most services offered seem to be identical.

Finally, it has to be pointed out, that the starting point for any systematic seek of *innovation* is the opportunities analysis. Depending on the nature of the innovation the same opportunities have different importance for different business segments. For example, certain demographic data could be of minor importance to the heavy iron industry. Analysis of the same data is on the contrary of major importance to industries producing industrial consumer products (i.e., electric home appliances, food products etc.)

4.1 The Coca Cola Adventure

It is worthwhile to quote here the dramatic decision of Coca Cola to change, after 100 years of indisputable leadership, the taste of its product, which had – in the meantime – been converted into a myth, and a symbol with a thousand connections and transactions.

The entire story started with the severe competitive activities noticed between the two major companies dominating the Cola-type refreshment market. When the turn overs involved are that high (it was then a market of 300 M$) the rules of the game are different. Even a minor market-shares re-allocation represents a certain danger. So Coca Cola, despite its leading position, felt insecure.

A very serious preparation procedure for the introduction of the new taste, has been followed. For almost 5 years the company contacted 190,000 blind taste tests which at a rate 2 to 1 showed a consumer preference to the new taste.

Early in May 1985 the effort to promote in the USA a new Cola with different taste started. But as soon as the new product replaced the traditional one, consumers reacted in an unexpected way. Within a few weeks hundreds of thousands of 'unhappy' customers kept calling Coca Cola asking for the old product they all knew and loved. Marketing people were driven crazy, they all lost their sleep, but they could find no way to convince all these people for the need of the change.

So, on June 10th of the same year (only a month later!!!) Coca Cola was forced to announce that the old Coke was soon going to be in the market again with the name Coca Cola Classic. The new taste was to be continued, but even this announcement did not stop 600,000 complaining telephone calls in the following three months!!!

So, what was all about? Did the major corporation make a bad brake instead of a well planned marketing movement? All facts lead to the conclusion that the experiment was completely unsuccessful despite the fact that more than 100 M$ were spent in advertising expenses.

The phenomenon was also analysed by the media. Many American and European magazines claimed that Coca Cola finally failed in convincing its customers for the need of change and thus was forced to take it back. According to 'Business Week' the case could be characterized as the public acceptance of an enormous failure in the marketing era. While others, more politely, referred to it as a quick response and adaptation to consumer's preferences.

Some market analysers did not hesitate to call it a well planned strategy which guaranteed Coca Cola a free publicity. The new Coca Cola has been developed according to them only in order to stop consumer movement to the Pepsi refreshment. Roberto Goizueta, the Coca Cola President at the time, was forced to declare that when his company introduced the new taste, did not intend to continue sales of the old Coke.

Even today, almost ten years later, marketers have not yet thrown light upon the case. The story is not 'hot', but the questions are still in front of us. Maybe the best explanation is the one given by the 'Financial Times' at the time of the crisis: Despite the money and the human resources spent in an enormous five years' market and consumer habits' analysis, it was not possible to trace the deep sentimental bond of thousands Americans with the old Coke.

5 The Innovation Musts

5.1 Simple and Impressive

For an innovation to be effective and commercially successful, it must be *simple* and *impressive*. It must perfectly answer a specific need. Otherwise it runs into the danger of confusing the average consumer, who generally hates long 'user's guides' and special difficulties.

The simplest effectiveness criterion, for a given innovation, is to create a first impression, that brings up questions like, 'But it's so simple! How didn't we ever think about that before?'

5.2 No Complicated Promises

An effective innovation must always be introduced in a simple way. It must not be *complicated*. Customer's satisfaction, for a certain need, should easily be recognized through the innovation. *Complicated promises*, for revolutionary applications, will most probably fail to excite the average consumer.

5.3 More Work than Brilliancy

It is not to be forgotten, that an innovation, despite the first impression, is always *more work than brilliancy!* There is no doubt that there are certain people talented in searching for the innovation. But *talent* is not the only ingredient of an effective innovation.

It is *talent, originality* and *knowledge* that can, at a first stage, guarantee an effective innovation. After that, what is also needed, is *strong and systematic work*. A businessman with strong *envisage* is also required in order to allocate the money and the resources needed to turn a brilliant innovation into a successful product!

References

1. Drucker, F. P.: The Discipline of Innovation. Harvard Business Review, May/June 1985
2. Drucker, F. P.: Management. Harper and Row Publishers
3. Kostoulas, G.: Planning. Management 2, Elliniki Evroekdotiki, Athens 1986
4. Levit, T.: Marketing Imagination. Harvard Business Review Bulletin, December 1983
5. Piercy, N., Giles, W.: Making SWOT Analysis Work. Business School and Strategic Marketing Development Unit, Marlow
6. Steiner, A. G.: Strategic Planning : What Every Manager Must-know. N.Y., Free Press 1989

7. Stevenson, H. H.: Defining Strengths and Weaknesses. Sloan Management Review
8. Tzortzakis, K., Tzortzakis, A.: Management. Athens 1992

A number of articles from both American (Business Week, Financial Times) and Greek (To Vima) newspapers and magazines have provided useful information.

Technological Education and Innovation in Connection with Tendencies of a World That Is Changing

Ilia Natali
International Institute for Technology Education
Villa Falconieri, I-00044 Frascati-Rome, Italy

Abstract. This communication including the ability to use languages effectively that both orally and writing are important, but today in this defined society 'of the image' that using innovative technology, results *extremely important* to know also the scientific-technological language: from mathematics to science and technology; from design to photograph; from printing and television; from video camera to computer; from informatic-language to computer and telemathique, etc. This knowledge helps the students: to discuss, observe, classify, measure, infer, hypothesise, experiment and create, etc.

Keywords. Computer, informatics, innovation, technology education

1 Introduction

In this cultural setting, where Prof. Blandow has summoned us to discuss about Technological Education, the seminar will be pleasant and profitable. For this I must really thank him.

I shall try to expound the fundamental points of such an important problem, clearly pointed out by the WOCATE Commission, through the general theme which introduces the most specific didactic-formative questions that I shall communicate later on.

I find it necessary to make at first a significant consideration: we are living a moment of the history of the world where the possibility of innovations, changes and movements, is extraordinary. For this reason the future societies will need people suitable prepared, always ready, above all to use their mind as well as their hands.

From such tendencies clearly emerges the importance of education and the necessity to conform one's own culture and professionalism towards the 'NEW' that is advancing with such a speed.

Undoubtedly the steady scientific-technologic development characterizes our time, and the Board of Directors appointed to the educational methods, have the responsibility to offer society a school able to answer the demanded requests.

Priority must be given to projects that are pertinent to the problem of 'continuity' in every order of school. Now it is applied between the maternity and the primary school; first approaches are being made between these and the 1st grade of the secondary school (Compulsary Education) but is also beginning between the above mentioned and the higher secondary school at an experimental level.

It is true, that in these last few years some *exciting innovations,* in particular with the introduction of the new programmes in the primary school have been finally introduced in Italy. The 'moduli' or (forms) in fact contain many capacities in the scientific-technological and the economical field, as they tend to introduce in the school, starting with children (aged from 6 to 10) a greater attention to the quality of education and in particular to the formative process in acquiring knowledge of a 'method to operate'.

But the problem is complex so it is important for the primary school teachers to be able to exchange ideas and experiences with the teachers of the 1st grade secondary school (who work with pupils aging from 11 to 14 years), to find the leading thread indispensable for the 'continuity' with the previous school, *in relation with the curriculum* to avoid useless repetitions, also regarding *the didactic-methodology and the standards of valuation,* considering the times of learning and the needs of the children according to their age.

1.1 Some Meditations

The problem of the education of man is thus assuming a primary role in every country of the world, above all for the radical changes, accured in the areas of social economy and technological human activities. Therefore we expect from man, a proper use of the innovative technologies, to be applied in the various sectors of the world of work.

Thus it is necessary to have a gradual serious education in the technological-scientific field, beginning from the primary school. We, who are working in the school, must collaborate with the families of the pupils, and we encourage the children to be aware of the role man has in determing the *quality of preparation.*
We can suggest proposals of actions tending to:

- develop creative capacities
- develop observative capacities
- develop methodological capacities
- develop handling capacities
- develop mental capacities
- develop adaptability to changes and
- consolidate the sense of responsibility even in view of different innovative activities

This obviously requires a more adequate training of the teachers, in every level of school, but this is not the centre for facing such a problem. The approach to informatics and multimedial technologies becomes then an object of study and at the same time an instrument of learning.

Never before in history the quality of life of the individual has been tied so directly to the technological innovations.

The students can learn history from technology applied to the computer and from its impact on today's society. We discover the effects of the computer applied to the world of production and information, but such objectives can be researched by remembering the period of learning and the age of the pupils. Let us then provide them with adequate instruments according to the principles of 'graduality'. A valid method of making the students aquainted with their formative process is based on didactive actions tending to educate to image and communication, all through the use of multimedial means, faster by learning to use the computer.

Therefore *a simple project* for learning to handle the elaborator could be the following, which was accomplished during the three years of the lower secondary school:

2 The Computer as Instrument of Communication

At this point our world is rapidly acquiring information and according to McLuhan we are starting towards a global village, where the fundamental 'media' is represented by informatics, that controls and runs everything in including communication.

This project has therefore the aim to face informatics as a technique; to analyse, manage, simulate and prevent the development of a real problem or situation.

In the basic professional training of a citizen the ability of reading and decodification of the quantity and the quality of 'information' and of 'communication' present in structures of every day use is imposed: from school records or filing-cabinets, to train time-tables, etc... A new alphabet must then take into account the new reality which would otherwise risk to be excluded from the school.

2.1 Informatics and Computer

Working at the computer is obviously the most important moment for learning techniques and knowledges, that will in any case be learned also in a theoretic stage and therefore more abstract.

We shall work on the utilization of interactive software because they require a direct relationship between the pupil and the computer: the same programme will in fact compel the student by proposing definitive operative way and so enable him to continue in the programme.

2.2 Hardware Structure

Another important moment for our students is that of the 'empiric' approach to this versatile and powerful machine that can receive an enormous amount of data and give back complex and elaborate information. Therefore it is important that the students learn the elementary mechanisms of the complex procedures of the computer, and be aware of its importance in the various fields of man's work.

It is up to the school to enable our children to use the electric and electronic circuits essential in informatics (logical circuits) and with the relative mental representations. We shall then be able to achieve simple circuits that point out the structure and the working of the logical operators, at first considered singly then connected among them to form more complex diagrams, that can simulate binary numbers.

Finally the students will be aware of the usefulness of this means even to verify statistic results for sample regarding environmental research, dealing with the changes of trades and professions, during a period of 10-20 years.

2.3 General Didactic Objectives

– Organization of one's own thought according to logics adequate for the needs of the pupils
– Comprehension of the individuality of informatics as a container of methods and useful instruments to face problems on the transversal plane
– Learning about contents or procedures, favouring education to organize one's own thought in a logic mental course to solve a problematic situation
– Enrichment of the vocabulary by using scientific terminology
– Comprehension of the structure and the working of the calculator to improve the scholastic curriculum
– Fulfilment of programmes by using the languages LOGO and TURBO-PASCAL

The learning of the LOGO is particularly significant from the pedagogical point of view, as it enables the students to acquire naturally an efficient method of work, that is valid for all disciplines.

The LOGO in fact, is not only a programming language, but above all it is an 'environment of knowledge' suited to increase the capacity to elaborate projects and to solve problems. The activities can be faced by the pupils with autonomy, and demand a first approach with pen and paper and subsequently to be checked at the calculator.

As for the PASCAL language, it has always been considered difficult or very binding to use, nevertheless the TURBO-PASCAL results interesting for many reasons: it is of general use, it is used for every type of application (numeric-sums, manipulation of the string of characters and graphics) resulting more powerful and faster than other progressive languages.

2.4 Disciplinary Areas Involved

- Technological-scientific area
- Literary area (Italian, Geography, History)

3 Articulation of the Project on Triennial Bases

First Class

Objectives

- Reflection on the machines and location of the automations
- Understanding (by concrete facts) What and elaboration is understood as series of operations, actions and changes that can transform an *Input* into a finished *Output*
- Decomposing a difficult task into simpler parts
- Drafting the resolutive algorithm of a problem with the projecting language and with the flux diagram
- Comprehension of the structure of programme, understood as a series of instructions
- Study of the LOGO programming language

Contents

– Symbology and construction of a flux diagram
– Differences and analogies between a flux diagram and a programme
– Adapting an adequate language
– Knowledge of the keyboard and the video
– Concept of the CODE as a way of communication
– *Codification* to formulate the message and *Decodification* as interpretation of the message
– Use of a WORD-PROCESSOR programme for the compilation of a text
– Compilation of programmes in LOGO relative to the Technical Education and Mathematics programme
– To analyse a multimedial environment problem starting from the search of data, and arriving at the solution in logically lined stage, through a resolving strategy

Instruments

- Specific software. Text. Monographs. Maps. Daily. Papers. Screens. Camera. Projector. Computer. Video-Camera.

Interested Disciplines

- Technological. Education. Mathematics. Literary studies (Italian, Geography, History).

Scientific Method

- Verifications

Compilation of a flux diagram and translation into LOGO of graphic, mathematics and statics programmes. Typewriting test with multiple choices.

4 Example of Didactic Unit (8 hours) 1st class

SLOGO (superlogo)

The Polygons

Objectives:

- learning to utilize the repeat control;
- learning to draw triangles, squares and regular polygons with the elaborate;
- understanding some properties of the polygons;
- carrying out decorative drawings.

The polygon is the part of a surface, limited by a closed breaking and not interwined. This means that the consecutive segments form a path without crossing. A polygon is regular when the sides and angles are congruent among them: it is thus equi-lateral and equi-angular. We can see that any polygon of n. sides is divided by diagonals coming from the same wertex in n.2 triangles (see figure). Therefore as the sum of the angles of a triangle is a flat angle, we can affirm that the sum of the interior angles of a polygon of n. sides is equal to a (n-2) flat angles.

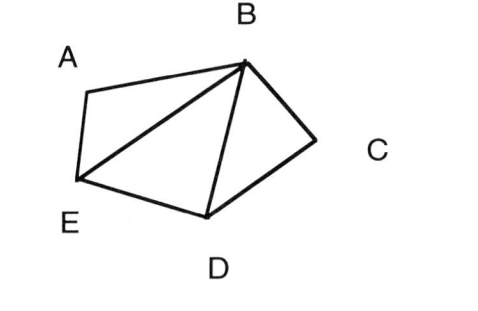

Fig. 1 Sum of Interior Angles=(N-2)* 180°

If the polygon is regular, to obtain the measure of an interior angle, it is sufficient to divide the result obtained by n.

To achieve the regular pentagon with the elaborator (sum of the interior angles =(5-2)*180°=540°) you must then divide 540 by 5, obtaining thus 180° and consider the supplementary angle (72°). Then you must write the following instructions:

A 100 D 72 A 100 D 72 A 100 D 72 A 100 D 72 A 100 D 72 A 100 D 72 A 100 D 72 A 100 D 72 A 100 D 72

The instructions A 100 D 72 can be considered 'a modulo' that is repeated 5 times and in the LOGO language exists a (Repeat) control that carries out the repeated instructions.

Repeat: number of repetitions (instructions to be repeated). Makes repeat the execution of a whole of instruction as often as required. It requires two arguments: the first is the number indicating how many times the repetition is necessary, the second in brackets.

To draw a regular pentagon you can strike: *Repeat* 5 (A 100 D72)

Second Class

Objectives

– Examination of the main parts of hardware
– Concept of (RAM-ROM) temporary and permanent memory
– Concept of CPU as Unit Logic Arithmetic and Control Unit
– Knowledge and use of MS-DOS Operative System
– Use of a Word Processor programme to be able to compose a text, an essay, a journal etc.
– Potentiation of expressive capacity by using multimedia
– Beginning to study Turbo-Pascal programming language

Contents

– Turbo-Pascal language: Editor's commands, Rescue, Structure of a programme, use of punctuation WRITE and READ commands, the VARIABLES
– Debugging of a programme
– System of binary numbers and electric circuits
– Compilation of programmes in Turbo-Pascal related to the programme of Technical-Education and Mathematics
– Analysis of climatic bands (Geography)
– Use of a programme related to human anatomy
– To get a deeper knowledge of the territory and about the use of the WORD PROCESSOR to write a text

Instruments

- Specific software

Interested Disciplines

- Technical-Education, Mathematics-Letters, Scientific Method

Verification

Compilation of flux diagram and translation into Turbo-Pascal of both graphic and mathematics programmes. Typewriting and relative printing of reports and news Comprehension test with multiple choice questions

Third Year Class

Objectives

– Deepening the knowledge of the Turbo-Pascal programming language
– Construction of an electric circuit corresponding to a binary algorithm
– Use of Basic Data to be able to gather up, to file and to elaborate data also for the purpose of production of gate graphics, aerograms, istograms, diagrams etc.
– Potentiation of expressive capacities on the transversal plane

Contents

- Turbo-Pascal Language: meaning. IF... THEN... ELSE...; WHILE...DO; logic operators AND and OR; meaning. REPEAT... UNTIL; FOR... TO... DO; GO...TO.
– Filing data understood as gathering the same type of information as the ones to be elaborated
– Gathering of data regarding library books, collection of video cassettes scholastic population etc.
– Learning of the graphic method and the capacity to compare it with the algebraic method
– Calculation of the probabilities
– Study of astronomy (Geography)
– Use of the electronic sheet (spreadsheet) and of the Basic Data
– Use of the orthographic corrector in a Word Processor
– Input of a graphic inside a text or in a report

Instruments

- Specific Software

Scientific Method

- Verification

Typewriting of a text and relative printing. Compilation of a flux diagram and compilation in Turbo-Pascal of both graphic and mathematical programmes. Comprehension test with multiple choice questions. Systematic research of data (e.g. book or video cassette) inserted in a Basic Data

References

1. AA-VV: La Scuola nella società tecnologica. Roma (ed.) Anicia, pp. 93-104, 1988
2. Fierli, M.: Guida alla tecnologia. Roma (eds.) Riuniti 1983
3. Popper, K. R.: Cogetture e confutazioni. Bologna-Il. Mulino, pp. 499-509, 1972
4. Brown, D.: Usiamo il cervello. Torino- SEI., Cap. III, 1976
5. Lariccia, G.: Le radici dell'informatica. Firenze- Sansoni (ed.), 1981
6. Andronico, A.: Scienza degli elaboratori. Bologna-Zanichelli, 1974
7. Ravaglioli, F.: In Rivista dell'Istruzione. L'Educazione Tecnologica. Rimini-Maggioli (ed.) Giugno, pp. 329-333, 1993
8. Pastic, M., Ketele, J. M.: Osservare le situazioni educative. Torino- SEI, 1993

Integration as an Innovation Course in the Education Concerning All Fields of Technology

Kazimierz Uzdzicki
Pedagogical University
Department of Technology Education
Al. Wojska Polskiego 69, 65325 Zielona Góra, Poland

Abstract. Technology as a subject being taught in secondary schools requires a special educational conception since it includes contents of different branches of technology. There are two possibilities of its realization. The first concerns educational contents. Its importance lies in the sequence concerning different fields such as wood, metal and plastic working and the following realization of each field separately.

The other conception, according to the author – a more rational one – consists of the integral approach which combines the teaching of the working and processing of the three above mentioned materials together. Activities such as cutting, drilling, sawing and turning, etc. are the essence of this conception. Hence the students are capable of a quicker transferring and distinguishing similarities and differences of the activities performed on different materials.

The article will comprise the possible and integrated solutions concerning all fields of technology education which is now being called 'technology' in Poland.

Keywords. Technology education, secondary schools, innovation

One of the trends of improving the educational process in the field of technology teaching at the secondary school level is the problem of integration in its vast sense. It is one of the main sources for the best realization solutions. It seems right to consider the research on integration to give many other possibilities for the good of the process connected with education concerning all fields of technology. The aim of this paper is to try to show the integration as one of the important tasks of the contemporary educational system at the primary school level referring it to the comprehensive school subject called 'Technology Education'.

The existing content structure of technology education in primary schools of a comprehensive kind (i.e., IV-VIII forms) is based on their selection according to the character and distinctive features of the respective materials. And so in form IV there is woodwork, in form V there are plastics and textile fibres, in class VI nutrition engineering, in VII metalworking and in VIII the elementary information about electrical engineering and electronics (1). While working on the improvement of the process of education, we could think about other arrangements of content. The integration approach is one of the novelties in education concerning all fields of technology. It requires a different way of thinking in at least two areas: a) in the content structure b) in the organization of pupils activities. The innovation includes searching for different specific teaching conceptions for technology education in secondary level schools. Integration as a phenomena lying at the base of improvement of the realization process of subject tasks, results from the fact that it consists of contents from various technology fields such as: materials science, process engineering, technical drawing, theory of machines, sewing, baking, the basis of electrical engineering, electronics and computer science, the problems of work organization, certain topics from engineering and management, elements of job introduction and others (2).

Taking into consideration certain confusion in the educational process including technology education which concerns mainly the overload of school programmes, certain studies have been undertaken to integrate the contents of technology education. Some of the chosen processes to integrate are: woodworking, metalworking and plastics treatment. The integration approach means that instead of the hitherto ways of skills training, i.e., separate in woodworking, separate in metalworking and plastics treatment, one integrated process of dealing with these materials has been introduced. This indicates that in the new approach the realization of the contents is done on the bases of one kind of activity but realized on all three types of material. For example: cutting or drilling is performed on all three types of materials: wood, metal and plastics. The content realization together with the activities includes the description of those materials, comparison of the properties at the time of treatment, grasping the similarities and differences, which is a cause for:

i. a higher activity of students
ii. the possibility of applying the problem approach
iii. a greater possibility of independent thinking
iv. the possibilities of linking the practical activities with the descriptive verbal act, expressing the similarities and contrast features.

The knowledge obtained in this way is more competent and the skills acquired by the students are long lasting and require less time to be developed.

The conducted studies in this area (3), as well as the practical experience of the introduction process of the new trend for realization of the three technologies, allow us to formulate the following conclusions:

1. The introduction of the integrated way of teaching technology to the programmes and syllabuses requires the teacher to change the way of dealing with the subject. It refers not only to the application of appropriate methods and

organization forms but also a different way of conducting lessons, creating situations for comparisons, creating generalizations and drawing definite conclusions.

2. The recently introduced operational system of teaching technology in schools, creates a need for a structural approach to the educational content. This means such a logical arrangement of the material, where each piece has its fixed place and fulfils certain functions, but also results from previously learned information which is at the same time the beginning point for the next pieces, thus creating certain kind of knowledge.

3. The structural way of dealing with technology includes the possibility and bases for transfer. It is evident from the transfer principle that content and operations mastered previously make the learning process easier for the further contents and have to be constructed so that the relations and interdependence between the ideas, rules and principles are shown. According to J. S. Bruner's thesis the structure of a subject requires such a way of content formulation so that " you can sensibly connect together many elements. Shortly speaking to learn the structure - means learning the way things are joint together"(4).

4. The necessity of integration realization of technology education is a consequence of the structuralism of knowledge importance. The integration can be looked upon in a few levels among others:

> I. H o r i z o n t a l - including the realization of syllabus in which on the bases of the shared theoretical regularities, the subsequent technological operation is applied to all three kinds of materials at the same time. This means, that if we deal with for example laying out, cutting, sawing or drilling, then we talk about it in relation to metal, wood and plastics. It is evident from our hitherto results that it is not the same which material we shall start the know-how teaching from.

> II. T h e   v e r t i c a l   i n t e g r a t i o n - belongs to the most difficult and comprises full structural technological content realization. The assumption of the vertical integration is the performing of specific technological operations on three kinds of main materials (metals, wood, plastics), in correctly arranged educational structures with common features. They include certain structures of activity in the framework of a certain operation group. These activities are the groundwork for comparisons and generalizations. For example if we teach about surface treatment, we deal with sawing, manual or machine planing, turning, milling and the finishing treatment.

5. The correlation between the particular rules, principles and laws of learning and skill systems should take place in both integration planes. However, in the second

one it has a little different character and requires the teacher to have a deeper educational knowledge.

6. The solution of the integration of the vertical or horizontal problem calls for improvement of the conditions to be able to accomplish the above mentioned proposition. We can name here among others: examining of the conception, improving the equipment base, getting acquainted with the new methods of teaching and an appropriate proportion between theory and practice.

7. The necessary integration condition for cognition and acting and also shaping the accurate kind of information is the problem of acquiring knowledge at lectures and laboratory classes. This particular method gives possibilities to compare and analyse the most representative phenomena and methods and thus be able to generalize (synthesis process) simultaneously taking advantage of the knowledge acquired earlier from such fields as material technology, chemistry, physics and others.

The mentioned problems concerning the content integration in the area of process technology do not fully deal with the elaborated issues. Many questions remain open to which we have to find answers. Integration does not only mean how to teach the skills for metal, wood or plastics treatment in the best possible way. The problem of students' awareness to the process is also interesting. A closer integration relationship can be seen when the appropriate dependence between theoretical content and practical activities take place. The teaching course requires a further study especially with reference to the vertical integration of technological contents.

It appears that further research on these issues will make it possible for us to acquire answers to the questions which have been put forward, the problems of which have not yet been elaborated up till now.

References

1. Primary school programme. Technology Education for IV -VIII Forms
(Program szkoly podstawowej. Technika dla klas IV-VIII) Warszawa, 1992
2. See p. 3
3. The Research on Integration Education of Some Technical Activities Conducted in the Institute of Technology in Pedagogical University in Zielona Góra under the supervision of the author of this paper in the years 1988 - 1992
4. Bruner, J. S.: The Educational Process (Proces ksztalcenia). pp. 97-98. Warszawa, 1974

Methodological Initiation Aspects of Educational Innovation

Maria Jakowicka
Pedagogical University
Department of Pedagogy
Al. Wojska Polskiego 69, 65-325 Zielona Góra, Poland

Abstract. Educational innovations arise on the basis of new tendencies in education which results from two sources. The first one is the development of science, among others psychology, the other one concerns social and economic changes. The innovations are a continuous process and thus should be examined in the process: the creation of certain innovation ideas and their practical application. There are different kinds of activities in each of the above mentioned stages. This paper embraces a short description of each of them.

The basic methodological problem asks for a more favourable output mechanism and a better application of innovations. This can be illustrated by the presentation of two approaches. The first one initiated by appropriate centres inspiring the teachers to experiment and the other expecting the teachers to undertake the initiative themselves.

In this paper there will be an attempt of a description and assessment of both methodological conceptions.

Keywords. Creativity, educational innovations, innovation methodology, reforms in education

In a complex and dynamically changing society it is necessary to design education in such a way as to be able to educate an active, creative and full of initiative man. He needs to understand himself and the world, he should be capable of organising his own performance, communicating with others, achieving more and making progress. It is not possible without introducing changes in the contemporary educational system. Initiating innovations is ranked among the main educational contemporary problems and is an indispensable factor to make the school and the teacher function. The innovations are a continuous process and should be considered in the categories of *continuity* and *modification*.

The improvement of practical realization on the basis of the results of sociological, educational and psychological studies, which are carried out because of the changes occurring in contemporary conditions of life and the development of children and teenagers should be a systematically worked out trend in the general professional activities of a teacher. The demand for changes increases at the time of educational reforms. Today, because of the occurring profound changes in attitudes, the innovations concerning the main educational assumptions require new teaching strategies and some detailed practical solutions. These changes inspire the need for constructive undertakings by teachers in designing and initiation of innovation solutions of educational methods, individual teaching programmes and a new organizational structure.

As it is known, at the period of reorganization the teacher has a double role to fulfil. On the one hand, he is the receiver of the innovations which result from the reform assumptions, on the other he is the author of the particular innovation solutions. Educational reforms are never designed in such a way as to stop the initiative of a teacher. Basing on the new principles, the teachers activities go together with their popularization and the innovation activity. Each of the innovation scopes needs a different activity range of creative character.

We can distinguish a *complete* change which takes place in all the elements of the system and a *partial* one - where the change occurs only in some elements. However in such a case it is necessary to verify and reflect upon the relations between the new change and the other elements of the system.

In the educational changes of the practical realization of the project it is essential to think creatively, i.e., be aware of the divergence among theoretical assumptions and the necessary change that should take place in the personality of pupils in connection with the educational practice. We should be aware of the change in the construction of the project its application and verification in practice.

The introduction of innovations needs a few essential matters to be mentioned:

1. The innovation conceptions originate in the process of *creative experience*. In this process there is no space for copying solutions which are well-known or described in educational literature and applied in practical situations. The solutions are modified or new but they consider the existing situation. At this point the imitation is eliminated; but the role of critical thinking, discussion ability and the confrontation of our own ideas with other ones are emphasised. In this process the awareness of *our own contribution* towards an improved performance is necessary, and this requires a good knowledge of theory and practice. Taking into consideration the above mentioned aspect we could distinguish three levels of novelty:

- *Application* of new pedagogical solutions which were described in literature but have not been hitherto applied in practice.
- A creative *modification* of propositions presented in literature including the latest scientific elaboration and the conditions of practical appliance.

- Our own concept of practical strategy of the task realization which would be *compiled by the author* on the basis of theoretical literature and the critical analysis of practical situations.

Three stages can be identified in the process of creative application:

- *Fundamental knowledge*, i.e., the state of knowledge of a given person before he starts investigation, invention activity directed onto a better realization of the task.
- *The phase of active study* when the thinking on the basis of literature takes place and the critical valuation occurs.
- *Reconstructed experience*, including new and creative solutions which are applied in practical situations together with the analysis of their significance. The transformation of the experiment results into opinions and concepts giving a beginning for a better performance thus improving the reality.

Innovation conceptions, the new propositions of improving the existing conditions need to be propagated. It is a necessary element of introducing of novelty into the existing reality (practice). The creative teachers' experiences depending how the changing ideas are designed - if they are performed by one person (e.g., teacher) or a group of specialists can result in a twofold process. In the first instance it is a *social* experience as it is team work and the conceptions are verbalised and written in forms of ideas and description. Then slowly put into practice, accepted by one person, become an *individual* experience. In the other case, i.e., constructing the conception by creative people, the experience at the beginning is individual, but gradually becomes public and customary. The popularisation of creative experiences is the main condition of development of innovation movement and it advances the teachers' awareness.

2. The following point can be expressed in a question concerning the need of inspirational centres which could guide the innovation activity of schools and their teachers. It could be asked here about the role of the theorists in those processes. In other words: who, the teachers or higher rank authorities, is more necessary in the stimulating changes in education?

At the very first stage the higher rank authorities assert a greater activity of specialists in given fields. Those are the theorists who are the designers of the changes, who also widespread it among teachers, and slowly make this idea popular. The teachers adapt the conceptions to their circumstances and formulate certain solutions. This trend is connected with educational administration which frequently causes certain forms of dependence and compulsion. In such circumstances teachers feel they are rather the executive not the creative power.

The mechanism proceeding from the ranks means giving the teachers full rights and freedom to apply the new solutions in practice according to the their own ideas. The authorship versions are contemplated by the teachers themselves. They are also initiated by them according to their own needs and then registered in methodological centres or taken to the headmasters of given schools. This brings about the response of a narrow circle of teachers but with a deeper self-evaluation and personal satisfaction trend.

We have to put forward the question about the merits of these mechanisms which of them are more effective for the pedagogical innovations?

There are conditions to try out both these experiments in Poland now. The first one was developed in the seventies (mainly from 1971) at that time the work on school reform was initiated. It was introduced (with many modifications) in 1977; the second reform started in 1989 when the political system changed. This second one has lasted till today.

In the first operation the Country Council of Pedagogical Development (together with the Polish Teachers' Association and the educational administration) were the main inspiration centres. The Centre focused its attention onto gathering creative teachers and educational science representatives to direct the work on innovation conceptions and to help initiating them into practice. There also existed regional councils, the duties of which was to stimulate and encourage to modernize school work.

In the second period (1989) there was a need to seek for other, different solutions caused by the political changes. The Educational Councils were closed. The undertakings and initiation of innovation conceptions were to be performed by teachers themselves. The teachers were obliged to elaborate so called authorship classes, syllabuses and methods of schools work etc. The meetings with the theorists were greatly limited but the creative role of teachers was approved of.

Presently the two approaches can be evaluated and some features can be distinguished. In the approach proceeding from the superior authority:

- a direct relation occurred between theoretical and practical innovation conceptions,
- consultations in the forms of seminars on the teachers' propositions were taking place at various times of the progress,
- a considerable number of teachers took part in the innovation processes,
- there were subjective reactions on the part of the teachers who thought it to be a sort of pressure on their choice of papers,
- the innovation conceptions were broader and more systematic with a better understanding of the topics.

Proceeding from the ranks the innovation stimulation which occurred was as follows:

- there was definitely more independence and invention on the teachers part,
- the handling of the educational theory was not very efficient, there was very little contact with the theorists,
- a fewer number of teachers were engaged in the educational reforms,
- the teachers had a more profound emotional relation with the innovation changes,

- the innovation conceptions are weaker, poorer, frequently referred only to one of the elements of the system, did not consider other elements, sometimes were simply pseudo novelties.

Comparing the features of the two approaches we have to indicate that the approach proceeding from the superior authorities has more intellectual value, the elaborated conceptions are more complete and competent whereas the approaches proceeding from the teachers, i.e., the lower ranks are better in the sphere of motivation, satisfaction, independence and the feeling of importance.

Innovation in Initial Teacher Training: An Analysis of Benefits, Costs and Resource-Related Opportunities

Paul Griffiths
University of Brighton
School of Education
Falmer, Brighton BN1 9PH, United Kingdom

Abstract. Recent legislative reforms to Initial Teacher Training (ITT) in England and Wales have resulted in secondary schools taking a more significant role in the initial training of teachers. The rationale for this development is seen by the government as a means of raising the quality of ITT and in turn raising the quality of pupils' education in schools.

This paper considers some of the 'costs' and 'benefits' of the policy and practical implications of this development for schools and universities. It provides a particular focus on factors influencing school policy decisions in relation to ITT and on some of the issues of significance for the initial training of technology teachers, and the role of school-based mentors.

Keywords. Initial Teacher Training (ITT), school/university partnership, technology, higher education, teacher competences

1 Introduction

There has been a growing involvement by schools in the teacher training process since the early 1980s when the Council for the Accreditation of Teacher Education (CATE) was established. Evaluation by Higher Education of the operation of courses of ITT, coupled with legislative changes, have led to a recognition of the importance of quality school experience in the development of student teachers' professional skills and understanding.

The most recent element of the reform programme for ITT began in 1992, influenced by the Centre for Policy Studies (Lawlor 1990), with legislation designed to decrease the influence of Higher Education in the training of teachers and transfer that responsibility to schools.

From September 1994, all secondary ITT courses, and by September 1996 all primary ITT courses, in England and Wales will have to meet new criteria which the government has now set in place - (Department for Education Circulars 9/92 and 14/93) and be subject to external inspection. These developments are commonly referred to as school-based courses. The particular features of the new approach are:

– significant proportion of school-based training in which schools take the lead
– a contractual partnership between selected schools and Higher Education
– schools must satisfy certain training partnership criteria to be regarded as suitable
– HE has responsibility for training teachers for their new role
– HE is responsible for the Quality Assurance of courses
– assessment of students will be on a competency based system.

During the last two years, a number of colleges and universities have been piloting new secondary age phase courses which meet the Government's reform criteria as set out in DfE Circular 9/92. The consequences of the reforms, particularly for the training of technology teachers, raise a number of issues relating to the management of this change. In the longer term, there are opportunities for the creation of even stronger professional partnerships between schools and universities relating to technology than have traditionally existed and which could enhance staff development and career opportunities for technology teachers.

This paper analyses some of the challenges, achievements and future potential of school/university partnership which have been a feature of recent developments. Some of these were anticipated and planned for, others have become clear after the evaluation of a pilot scheme. It is perhaps possible to consider these issues at:

i) a macro level - relating to whole school policy approaches to student teacher training; and

ii) subject specific level where there are distinctive issues which relate to the training of, in this case, technology student teachers.

The statutory minimum range of competences student teachers must demonstrate before they are eligible for qualified teacher status are defined in Department for Education Circular 9/92. These competences are grouped under the following headings:

– Subject knowledge
– Subject application
– Class management
– Assessment and recording pupils' progress
– Further professional development.

Additionally, there is a range of broader professional qualities which extend beyond this framework and which students would be expected to acquire to be judged ready to teach.

Within and beyond this range of competences, it is possible to differentiate two broad elements to the school-based training:

1. Generally applicable professional issues;
2. Subject specific related competence.

Developing these competences has become the joint responsibility of schools and universities and a pilot project run during 1993/94 has been identifying the respective roles and responsibilities of schools and universities within the new structure.

The range of student competences required make it unlikely that a single member of school staff will have the necessary background and expertise to take sole responsibility for training students. In conjunction with partnership schools, these two principal elements of school-based training are formally recognized as being the responsibility of the professional tutor and subject mentor respectively. The professional tutor is generally a senior member of staff (Deputy Headteacher) and the subject mentor an experienced subject specialist teacher, though not necessarily a Head of Department.

These roles are essentially complementary, but the former is particularly significant in informing the school's approach and attitude towards teacher training through the link with the school's senior management team. Shaw (1992) identifies the importance of whole school management issues in formulating teacher training as a framework for the professional development of their staff. In many cases, this has entailed a cultural change for schools and apart from their role in providing elements of the student training, the professional tutor is a key agent in effecting this change.

From the initial stages of planning, it was important to acknowledge that the new approach would not be successful if it did not recognize that, as Edwards (1990) has put it, partnership between schools and colleges is not simply a matter of adding more of the same, but bringing different, complementary strengths together. The differences are both quantitative and qualitative in nature.

This meant the new developments needed to support the change of role for schools and teachers in moving from a form of guardianship, as Wilkin (1992) has described schools' traditional role, to a role which involves a major professional responsibility for the students in training. However, before embarking on partnership, there are a number of policy issues to be determined by schools.

2 Whole School Policy Issues

Amongst the issues which relate particularly to decisions by school senior management and governors are:

1. The implications for the school culture of significant involvement in ITT;
2. The priority of schools to educate pupils;

3. Priority of duties for senior staff;
4. Quality implications of student training;
5. Adequate resourcing and transfer of funds to school;
6. Providing time for school staff to train for and perform new roles.

Although schools have always contributed to the teacher training of students, taking significant responsibility for Initial Teacher Training has not traditionally been part of their culture. A report by HMI on School Based ITT in 1991 confirms that headteachers and governors see the principal purpose of schools as teaching pupils rather than training students. There has been a growing awareness by schools that partnership carries with it a significant responsibility for course delivery, which extends to a qualitative change of emphasis from the sort of support traditionally offered to students. This new role entails not only providing formal critical guidance on students' practice in the classroom, but structuring provision for students to experience a range of school experiences and an introduction to the basis for school policy in certain areas. Also, the assessment of the students - an element of the role with which teachers are least comfortable - has become a new expectation. Thus, as identified by McIntyre et al (1994) and confirmed widely by teachers themselves, teachers' experience of school-based training is complex and demanding.

Consequently, there has been ambivalence on the part of some schools towards commitment necessary for full involvement in partnership training. However, amongst the positive aspects of involvement for schools, students are seen to often bring new ideas and enrich curriculum discussion and wider educational debate in the school. In the case of technology, a subject area characterized by a high rate of change, students are often seen as important 'change agents' contributing a new area of subject expertise.

Additionally, there has been recognition by many schools and teachers that central involvement in the training process can cause teachers themselves to benefit by reflecting more analytically on the most effective practice. Training can also provide a further professional dimension for senior staff and subject staff with increased job satisfaction. It also formalises a link with the universities where school staff and university staff are brought together for mentor/professional tutor development sessions. In some cases, there can be accreditation of mentorship training for individual staff. This might comprise part of a diploma or higher degree award. This is one means of formally recognising the importance of those involved in the training role. Bailey and Brankin (1992) in a case study of a technology mentor, confirm the importance of according the mentor role status within the school management structure. Thus, there are a range of benefits to schools from involvement in a partnership training process.

However, there are quality-related issues which have caused schools concern. One quality issue relates to the proportion of their curriculum time that pupils receive teaching from students. Published league tables of schools which reveal pupils' examination successes put increasing pressure on schools to score highly on this particular performance indicator. Student teachers can lack the experience necessary to make a strong teaching contribution. Consequently, in an era of increasing accountability to school governors and parents, the number of students undertaking training in school at any given time is a quality-sensitive issue.

The effect of students on the school's level of quality provision and the diversion, particularly of senior staff, away from the immediate needs of pupils towards an involvement with teacher training, has raised, for some schools, concern over reconciling these demands on staff time. These factors have led to some schools withdrawing teacher training placements. The loss of quality departments and schools from the teacher training system make it increasingly difficult, if not impossible, for the university to exercise any significant degree of discrimination in the choice of schools with which it forms training partnerships.

However, in addition to the above benefits, fortunately many schools recognize the value of the wide practical experience of commerce and industry possessed by mature students who enter technology teaching, coupled with, in many cases, strong subject expertise, can mean they make a particularly valuable contribution to the school during the training period. Additionally, a number of schools have subsequently recruited successful students to the staff of the school, having had the opportunity to evaluate the student's ability first hand.

Resource Transfer to Schools. The recent financial autonomy of schools has resulted in their increased awareness of budgetary control and revenue from external sources. Determining agreement on the level of funding to be transferred to schools has been the subject of much sensitive negotiation. It is salutary to note that assuming a PGCE intake of 150 students and a resource transfer of £1000 per student is the equivalent to approximately 5 university staff FTE. The consequences of this, particularly for small HEI institutions or departments is obvious. Indeed, a small number of university departments have recently withdrawn from secondary teaching training. Without the funding to sustain a critical mass of HE-based staff, the institution's ability to maintain a quality contribution to ITT, along with in-service training of teachers, and research becomes threatened. The problem is exacerbated by SCITT courses receiving higher funding levels than those HE based. Graham's recent account (1994) provides a detailed analysis of resource implications of school-based ITT.

However, by developing a school/university partnership which conceptualizes the nature of that 'partnership' as broadly defined and extending beyond initial teacher training is providing a clear role for the university in contributing to the school's staff development policy and providing varied opportunities for school staff development. This can range from short course attendance to Masters or doctorate level work, together with individual subject-based consultancy provided by HEI technology specialists working alongside school staff. Formalising this as an element of a partnership agreement enables schools to receive agreed whole school staff development support as part of a contractual arrangement on financially preferential terms as partners, thereby consolidating the school-university links and providing partnership which offers practical benefits to both school and university. Future development of this concept provides a major resource related opportunity for both schools and universities.

3 Subject Specific Mentoring by Technology Teachers

Analysis of technology-based mentorship indicates a range of issues affecting the availability and the role of mentors at subject specific level, some of which are perhaps uniquely associated with technology teacher training. These include:
- specialist trained teacher shortage in technology;
- responding to the diverse range of technology ITT courses;
- the move from a support role to a training role;
- the need to enable students to extend their subject knowledge;
- providing safety guidance/training in workshop procedures;
- responsibility for assessment of students;
- compensating for absence of student peer group interaction.

Within these distinctive areas, the technology mentor is supporting and guiding students' practical teaching, providing practical and tutorial guidance on theoretical aspects of the teaching role, and contributing to ITT course planning and evaluation.

Mentor Shortage. As indicated earlier, identifying sufficient specialist departments and mentors can be problematic. Technology has, until this year, been formally designated as a subject where demand for teachers exceeded supply. The consequence of this for school-based training has meant a limited number of suitably qualified and experienced staff to act as mentors. There has been a pool of unqualified teachers of the subject known as 'hidden vacancies' making the number of specialist qualified teachers appear greater than it is in actuality.

Mentors generally have found adjusting from the role of teacher support to teacher trainer a particular challenge. Because teachers' work is principally directed towards pupils, good teachers cannot necessarily automatically define and share with students the constituents of good practice, and this too involves a developed skill evolved in conjunction with the university.

The actual number of appropriately qualified specialist technology teachers is further confused due to DfE projected teacher numbers failing to differentiate between numbers required for the subjects contributing to National Curriculum Technology, viz. Design & Technology, Home Economics and Business Studies; thus apparently adequate recruitment of specialist teachers in reality results in inadequate supply, normally in the area of Design & Technology.

Diversity of Training Courses in Technology. In 1988, the HMI Report 'The New Teacher in School' found that for the secondary age phase, the 1-year full-time PGCE course was the principal route to gaining qualified teacher status for many subjects. However, in the case of technology, a variety of alternative routes have been designed by higher education to attract not only graduates but industrially qualified non-graduates into teaching. This has been partly motivated by a desire to increase the supply of secondary technology teachers by drawing on as wide a range of students as possible, but also in recognition of the quality to the profession provided by mature students with a wide range of life and industrial or commercial experience. This view is confirmed by the recent research of Adamson and Kennard (1992). The percentage recruited to teaching technology for 1993/94 is indicated below:

Table 1 National Balance of Undergraduate & Postgraduate ITT Intake for Technology 1993/94

Subject	Undergraduate	Postgraduate	Total
Technology	1,232 (55%)	987 (45%)	2,219

Each course will have a number of distinctive features and students are required to spend varying lengths of their course in school. These are:

		Total course length	Time in school	% of course in school
PGCE	1 - year full time	36 weeks	24 weeks	66%
PGCE	18 months part-time	36 weeks	18 weeks	50%
BA (2-year)		2 years	24 weeks	37%
BA/BEd (Hons) (4-year)		4 years	32 weeks	25%

Furthermore, this school-based period will be organized in a variety of ways which includes serial attachment, and block placement at different points through the year.

Catering for this diversity has put pressure on schools at a time when they are gradually adjusting to their new roles in teacher training. Managing, sometimes simultaneously, the various needs of students following different courses, who appear in schools at different times of the year and with different attendance affecting the mentor's role in school-based courses.

Breadth of Subject Competence. Graduates training to teach National Curriculum Technology are required to possess knowledge and skill in areas often outside those which formed part of their undergraduate studies. Most graduates entering ITT courses are normally qualified in only one of the contributory areas of technology - for example, electronics, or industrial design, maybe architecture etc. Thus, there is an immediate dual challenge for the student teacher of technology: to not only acquire the same professional skills as his or her counterparts in other subjects, but also to broaden their range of subject expertise in order to adequately teach the subject as defined at school level, together with relevant safety training. This means subject mentors arranging an individual subject enhancement programme for students. Our experience confirms recent research by Abbott and Evans (1994) which identified concern of students, teachers and lecturers that technology students are not developing a sufficiently broad range of practical subject competence.

This problem has been compounded by the relative isolation of technology students from their contemporaries whilst in school. Traditionally, technology students have benefited significantly from mutual support provided by university-based peer group interaction, where they have been able to exchange subject knowledge and skills and undertake comparative evaluation of their developing professional subject competence during university-based sessions. School-based courses make this more difficult to achieve with schools, for reasons already identified, often reluctant to take more than one subject specialist student at a time.

4 Conclusion

Recent innovation in school-based technology teacher training has the potential to provide a range of benefits to both partners in the training process. However, these benefits will only be realized and training standards increased when the following issues are resolved. These include:

– ensuring an adequate supply of school departments which are regarded as good practice role models for the technology student;
– similarly, maintaining an adequate supply of qualified technology mentors;
– schools and departments withdrawing from school-based training due to concerns that the quality of their pupils' learning will be reduced;
– pressure from schools to increase the payment they receive but sustaining financially viable university provision with a critical mass of staff to support ITT inset and research;
– enabling technology students to broaden their range of subject knowledge and skills and undertake necessary safety training whilst in school;
– developing a quality assurance system which can effectively monitor teacher training, up to 66% of which is off campus in a large number of schools covering a wide geographical area.

However, there is evidence that schools can benefit in a number of ways, particularly when they recognise ITT as part of the school culture and, when this is reflected in a clear, whole school policy commitment, in these circumstances:

– technology mentors have an additional professional dimension to their role and increased career development opportunities;
– broadly-based partnerships between university and schools which extend beyond

ITT and formally recognize a contribution the university can make to whole school staff development in return for the schools' involvement in training (potentially particularly beneficial for technology);

Additionally, the presence of technology students can:

- enrich curriculum development and subject debate and result in teacher reflecting on professional practice;
- mature students with commercial or industrial backgrounds offer a wide range of relevant skills and experience to schools during their training period.

It will be a further issue to undertake a detailed comparative analysis to determine whether school-based ITT results in the government's intention to raise training standards. On the limited evidence to date, this is not necessarily the case.

References

1. Abbott, I. D., Evans, L.: Separating the HEAP from the SAP: Initial Teacher Education in the secondary PGCE sector. In: Mentoring. Vol. 1 (3), 1994
2. Adamson, F., Kennard, R.: A Study of Mature Entrants to the Teaching Profession. HMSO, London, 1992
3. Bailey, T., Brankin, M.: Establishing Criteria for Mentoring. In: Wilkin, M. (ed), Mentoring in Schools. Kogan Page, 1992
4. Department for Education: The Education Bill: An act to make provision about teacher training and related matters. HMSO, London, 1993
5. Department for Education: Circular 9/92 Initial Teacher Training (Secondary Phase). Department for Education, London, 1992
6. Department of Education & Science: The New Teacher in School: a Survey by HMI in England & Wales. HMSO, London, 1988
7. Department of Education & Science: School Based Initial Teacher Training in England & Wales. HMSO, London, 1991
8. Edwards, T.: Schools of Education - their work and their future. In: Thomas, J. B., British Universities and Teacher Education - A Century of Change. Falmer Press, 1990
9. Graham, J.: Funding Issues in Establishing the Teacher Training Agency. Paper to the UCET Management Forum. United Kingdom, 1994
10. Lawlor, S.: Teachers Mistaught - Training in Theories or Education in Subjects? Policy Study 116, Centre for Policy Studies, 1990
11. McIntyre, D., Hagger, H., Burn, K.: The Management of Student Teachers' Learning. Kogan Page 1994
12. Shaw, R.: Teacher Training in Secondary Schools. Kogan Page, 1992
13. Office for Standards in Education: Working Papers for the Inspection of Secondary Initial Teacher Training. 1993

Technology Education, Innovation and Management

Alzbeta Dingová
CHIRANA-PREMA a.s. Stará Tura
Factory for Medical Instruments
Námestie Dr. Alberta Schweitzera 194, SK-91601 Stará Turá, Slovak Republic

Abstract. CHIRANA-PREMA a.s. in its holding arrangement is the leading manufacturer of medical and measuring equipment in the Slovak Republic. The structure of production programme and development of business activities build on almost 60-year tradition of engineering industry in the town of Stará Turá. This production was constituted as a concern during the period of the Czech and Slovak Federal Republic and comprised major enterprises for production of medical equipment in the whole former Czech and Slovak Federal Republic.

Keywords. Company, educational activities, management

As far as organizational structure is concerned, CHIRANA-PREMA a.s. has been transformed into the company of modern type. By making use of potential of whole joint-stock company through daughter companies with capital participation of CHIRANA-PREMA a.s., it ensures a wide range of activities concerning through development, production, services to public relations and sales.

1993, the foundation year of our denationalized company was not very favourable due to insufficient level of working capital, drop in sales on our major markets, lack of foreign capital, inadaptability of major part of employees to the conditions of market mechanism, etc.

Nevertheless, positive effects of internal restructure became effective. These effects consist of high level of decentralization, empowering more responsibilities to lower unit organs and above all in adopting marketing mix management methods.

But most important of all is the fact that firm assumptions were created for a long-term improvement of the current economic situation of the company and its future prosperity. This is from drafted strategic aims in main areas through the establishment of long term contacts in a good number of countries, to the decisive start of implementation of modern principles of management, among others, for example, quality management in accordance with ISO 9000 Series International

Standards, extensive system of continuous training of all employees, marketing management methods, and the like.

Accepted long-term educational programme and the development of management, marketing, administration, and finance, modelled on the free market economy conditions creates guarantee of future for our customers, employees, and shareholders.

Elaborating of our programme of education was based on analysis of our company. This analysis showed series of weak points and related dangers that could be effective due to insufficient qualification of our employees. These deficiencies are coupled mainly with the fields requiring new ways of training of thought as consequence for transformation of the whole economy to market principles as well as internal restructure that supports responsibility and decentralized decisions.

Common sign of these changed conditions is the fact that the whole firm is subordinative to the customer not to the tasks stated by higher authorities, as it was in the past. Analysis of joint-stock company showed the following weak points related more or less to qualification and professional abilities of the employees:

- bad habits of employees and managers, pertaining from the era of central management system, preference of volume ratios and priority of production,
- inertia of pseudo-social feeling - this also originates from the era of socialist equality, social order that prefers social aspects to economic aspects and is marked by false comradely spirit and social hypocrisy,
- lack of discipline and lasting euphoria of liberty that pass into almost anarchy and formation of partial disloyal policies - this factor is due to the change of social conditions but in the same time it is also due to lasting departmental interests and related buck-passing of the past,
- low level of innovation thinking arising from accustomed egalitarianism and suppression of new thoughts but also from cautious attitude to new thoughts, being logical defensive mechanism against central management system,
- insufficient language knowledge,
- insufficient experiences with doing business in the full-market mechanism,
- incompleteness of the system of quality and insufficient positive attitude of their employees to the quality of their own work as well as to quality of their surroundings,
- unfinished information system.

A certain part of deficiencies listed are, in general, of subjective nature from the point of view of individual colleagues but the other part follows from the lack of information and/or knowledge and experiences. Each of the above mentioned points can be changed only by a systematic influence on the people by means of formation of public thinking (through public relations), by example, repression as well as by purposive training of all employees with the aim to influence their mind and behaviour.

The connection of phenomena such as information, knowledge, experiences as precondition of working output brought us to conclusions that 'only the one who knows how, when, where and what is to be done in organization, workplace, he is able to identify himself with the work' and 'only when the employee obtains enough information, knowledge and positive thinking, he will be interested in the development of the work'. In this connection mutual influence of information and communication is essential. The question of information is viewed from two positions:

- flow of information within the organization in both horizontal and vertical positions, which is ensured by means of hierarchic management and outputs of computer technology,
- information obtained by means of educating employees namely in the form of supplementing the education and making it deeper.

The question of communication is being closely connected with the personality traits of the employee thus being the base of good co-operation in the organization and, in the same time, of adequate presentation of the company with regard to the customer. In the field of education the level of communication of employee is connected with his social competencies. A good businessman should know where he wants his management to go and because motivation and education are closely interconnected processes that cannot be done casually, it was necessary to solve them through a system. Before beginning the formation of the system of education, the management investigated some base questions:

- the attitude of the management to the education and preferring educational activities by their workers,
- how the employees understand the educational programme, their attitude to it, the extension to which they understand the education as their obligation or in their interest,
- how to arouse motivation to education and on the other hand sanctions against benevolence to educational activities.

Answers to the above questions stimulated the management to draw up the following measures ensuring successful implementation of the programme of education:

- to convey an attitude of the top management to the education and their intention to create the system of education through suitable information channels,
- to motivate staff members in order to be involved in successful course of realization of the programme through personal contact of the creators of the programme (Personal Department Staff),
- to make out models of learned managers of individual management sections (top management, technical management, production management, commercial management) to help executive workers in creation of educational activities for the individual,

- to state principles necessary in selection of educational activities for the individual within the programme being realized,
- to make out a system of recording of passed educational activities concerning individuals,
- to formulate procedures enabling the organization to make use of obtained information for the work of employees.

With these aims in the mind the system of education of the employees was developed. The system assumes:

- the efficiency and rationalism of education,
- systematic approach,
- periodicity,
- interconnection of educational process to other activities of the strategic management of the activities of the Personnel Department, for example career, remuneration, social evaluation and the like.

The system is oriented to:

- the preparation of young people for the work in the company,
- the requalification of the employees of the company to the new conditions of the work, new products and new professions,
- the establishment of firm knowledge and professional abilities,
- the extension of qualification for the performance of existing profession,
- the improvement of the qualification for self-assertion on the higher level of management.

The system is carried out by means of:

- external education - professional and language at educational institutions both in our country and abroad or through study stays in other companies,
- internal education - through the School of Managers,
- systematic training directly at the workplace under the charge of experienced instructors,
- purposive self-education,
- periodical adaptation - by performing other work within the company.

Educational activities are generally directed to:

- the adoption of professional knowledge and competencies,
- the adoption of social competencies, mainly from the point of view of managing interpersonal relations, creativity, assertion, team cooperation and independence in responsibility and competence,
- the adoption of appropriate methodological procedures of staff members to rationalism and high quality.

The system of education should involve all employees, naturally, with high rate of selection of the influence on the individual levels. From this point of view it can be defined common aims of education for all employees:

- attitudes and procedures of quality work in accordance with ISO 9000 series,
- general attitudes towards the customer,
- the adoption of the Codex of the Employee of the Company and its aims.

A part of the educational programme will be carried out as integral part of working process - this is mainly true in the case of general educational activities, establishment of skills for the performance of existing professions as well as skills important for the achievement of the aims of the organization.

Part of the cost of educational activities supporting individual professional growth as the condition of professional career will be covered by the company but it is assumed individual commitment of the employee either by time or finance. Participation in the educational programme is differentiated as follows:

- integral part of working obligations,
- voluntary as the criteria of the comprehensive evaluation of the professional growth,
- recommended for the maintenance of the occupation.

The schedule of participation in individual types of education according to recommended schemes of educational level in various time cycles stated for individual working level are planned by a direct staff member who is responsible for the participation of individual employees in the planned educational activities. This element of the activity of managing staff forms, at the same time, one of the criteria of their evaluation. The system of education should obtain general support by the management but also by other employees and the trade-union so that the participation in it be conscious and active.

The individual programme of education is elaborated according to above the mentioned categories for employees and potential employees in a hierarchic manner from the point of view of categorization of obligations, voluntariness and/or recommendations. Except for the compulsory range of education for which it is necessary to elaborate firm plans according to the individual workplaces, the approach to other forms of education will be on mutual basis, i.e., on the one hand the programme of educational activities will be elaborated according to individual group of employees, on the other hand the chiefs concerned come to an agreement with employees concerned about their participation in individual actions in order that required or recommended knowledge has to be obtained within given time interval (2 to 3 years).

Further non-standard forms of education are individually stated mainly for employees for whom further professional growth is assumed. The programme of education of these employees is worked out by a staff member who is by 2 to 3 level higher in co-operation with direct chief of the employee concerned. The programme of education of top managers, i.e., expert managers, managers of companies and plants and their deputies is elaborated in cooperation of the top management and Personnel Department by the General Director. Execution of the educational activities stated in this manner and approved by the employee concerned becomes an integral part of working obligations and evaluation.

With the aim to achieve maximal efficiency of the educational system there are appointed professional guarantors from the members of top management of the company for individual educational programmes.

Simultaneously with the improvement of working team by increasing its professional potential, great attention is drawn in the joint-stock company to work up a system of the work for the Personnel Department. The Department has the possibilities to influence the quality by selection of the persons showing interest in work in our company through selective sessions. Different approach is applied to:

- school-leavers and graduates,
- workers migrating within the joint-stock company,
- persons interested in work from other organizations.

The management prefers the university graduates as the future source of professional workers. To fulfill the aim to achieve necessary number of the workers from this source, they are given attention from the point of view of :

- the possibility of long-term evaluation of their interest in firm (co-operation takes 1 to 3 years)
- their own activity which can be shown by:

- working out the holiday tasks prepared by joint-stock company,
- study stays at the workplaces of joint-stock company,
- study stays abroad ,
- their own commitment during the meetings with General Director,

- long-term evaluation of personality assumptions followed because of the many-year personal contact with the workers of personnel and expert departments,
- evaluation of professional assumptions, carried out during the adaptation process in which the graduate is watched by the workers of personnel and expert departments and his expectations are being aligned with the requirements of the work.

This process is stimulated by the general knowledge that "it is necessary to bring up the experts according to the conditions of our company in reference to their permanent lack."

Are We Making Technology Education Attractive to Our Students?

John Eggleston
Warwick University
Department of Education
Westwood, Coventry CV4 7AL, United Kingdom

Abstract. Governments in virtually every country desire more technology education for economic and social reasons. Yet there is little sign that the technology education provided, (despite many successes) is attracting widespread student support or commitment - either for career planning or participation in schools. This paper reviews some of the reasons for the present state and examines teaching, industry and student attitudes. This is followed by some suggestions for development and an emphasis on the urgency for remedial action.

Keywords. Technology education, teaching, industry, teacher and student attitudes

1 Introduction

I wish to commence with three propositions, all of which seem self-evident:

1. Every society and every government wants more technology education because it is seen to be the key to a developed economy and to growth in national income.

2. Every individual wants more of the products of technology for their personal satisfaction, security, comfort, leisure and entertainment. The goods that consumers in every town and city desire are remarkably similar.

3. In consequence every education system is trying to develop technology education from the early years through to higher and post graduate education for boys and girls.

So why isn't it working? Why are there empty places in most degree courses in most countries? Why are many children switching out of technology in many schools and switching into subjects they find more attractive? In France the under use and low capability of the university and technology institutes (IUTs) has led to major, even desperate, attempts by Francoss Fillons, the higher education minister, to enhance their attractiveness.

One of our most cherished beliefs is that almost all students enjoy technology, eagerly await technology sessions and work enthusiastically therein.

In my work I visit many technology classes in many schools. Many projects are good – but many are uninteresting and only modest in their results. I watch the faces of the students, many are involved – but many more are dead. Alas it is the same in science, maths and most other subjects. But why is it so in technology I ask?

Kimbell of Goldsmiths' College, London (1994) has explored the matter. In a paper yet to be published he has identified three categories of involvement in technology:

- Motoring (active learning);
- Podling (non-committed learning);
- Static (inactive).

In the best lessons motoring was dominant for a maximum of 30% of the time. In most, podling was the norm – an average of 75%. In many there was a disturbing amount of static behaviour, Shield (1995) in this volume, offers confirmation.

In our search for answers let us be clear from the outset that human capability in technology, as in most other areas of knowledge, is for all practical purposes, unlimited. I can give many examples. When I have the tyres changed on my car the young worker at the tyre depot has no pretensions about being a technologist. His school record in technology was minimal, but with the aid of his pocket calculator and reference books, he can work out the specific requirements for the tyres I need. He can deduct the discounts, the special offer terms and the association membership concession, add on the tax and get the price right in seconds. He can operate the technology to fit and service the tyres with total reliability – also with impressive speed.

Young people who appear to has little capability in school technology have amazing ability to tune motor cars and motorbikes and know exactly how to enhance performance. They are able to develop hi-fi systems, using sophisticated science and technology capability.

One of the most exciting things in my career is to be Chairman of the Judges of the Young Electronic Designer Competition, a major national competition in Britain, sponsored by the Texas Instruments and Cable and Wireless companies. Entries come from children all over Britain, and the candidates' capability to devise useful, practical electronic devices is breathtaking. What is so astonishing is how few of the entries are coming through the schools. Most of the candidates learn about the competition through the electronic magazines and the components manufacturers bulletins. We invite their teachers along to the finals and candidates often astonish their teachers by what they have been able to achieve. A very common remark from teachers is, 'I never knew that boy or that girl had the capability to do this'. Children have a secret world of achievement and capability that we as teachers very frequently do not recognize. Let me emphasize that I am talking about girls as well as boys, many of the successful candidates in the Young Electronic Designers competition, and in many similar competitions, are girls.

What can we do about it? We have the need and the human potential, why can't we get two things together? I think there are three areas to which we have to attend. They are ones that we are all very familiar with – teachers, industry and students. Those are the three raw materials with which we have to work.

2 The Role of Teachers

Let us look at the teachers first. We have a major problem in many, many countries. First of all there is difficulty in getting able students into teacher training courses for technology. As Cotton (1995) in this volume confirms: 'In many countries the most able students are tracked into pre-science and other 'high status' subjects and discouraged from technology and 'practical' studies.' That means that there is a relative shortage of graduates, and particularly good graduates, in this subject in many countries.

Secondly, the demand for those technology graduates is growing from industry, and industry, commonly, is able to provide a more attractive salary and incentives than the teaching profession. So therefore the supply of good, interesting, lively, graduates to technology teaching is often very limited. It is a vicious circle, because that means we are less likely to produce many good school technology students and so we enter a declining spiral.

At the university and polytechnic level we can often recruit highly able lecturers in technology, but there is a very similar problem. Once one gets a post in a university or polytechnic institution the emphasis in not on good teaching but on research. I speak as the person who chaired the teaching methods programme at my University for many years, trying very hard to ensure good teaching as well as good research. But at the end of the day, promotion is almost wholly determined on research and publication, teaching is one of the minor criteria. At a university I am familiar with (not my own) one notorious lecturer who delivered an essential part of one of the technology courses. Students were obliged to attend because it was a core part of the syllabus. But their boredom was so great that they did not only carve their names on the lecture room tables, they also amused themselves by floating paper aeroplanes and launching pellets at the lecturer. He was totally indifferent, as he stood writing away at the chalk board, not noticing what was taking place behind him. The university used to have two cleaners standing by to clean up the lecture room after each of his performances. In no time at all he shot through the ranks and was appointed full professor because his research was brilliant. He is now chairing a university department and the teaching of his department, I am sure, is not one of his major concerns. And so in the university system too, the fall out rate of students in technology is high and we are losing, yet again, more of the potentially able people who could deliver effective technology education.

There are other major problems about technology teaching. At my university, we make technology an obligatory course for all our teacher training students. But that is because of our enthusiasm, not because it is required by law. Many teachers are, frankly, scared of teaching technology. A recent survey taken of British teachers, conducted by Wragg (1989) found that something like 70% of all primary school teachers were alarmed and felt ill equipped to teach technology even though, by law, they have to do so. Even more to the point, they have great difficulty when they do actually teach technology to make it interesting, attractive and involving.

Many teachers are making real efforts to engage children by involving them in decision-making, helping them to think creatively to solve problems rather than simply getting the right answer. But many teachers still give higher grades to children who get the right answers, rather than those who try to achieve an interesting and original way of working. For many teachers it is hugely important to get the right results, I have seen many children working hard to devise a way of solving a problem, only to find that it is not the way that delivers the right text book result or the right combination of processes and so they achieve only low grades. The most able children find it boring and unrewarding.

Why do you think technicians are in schools? If we are perceptive we may agree they are there not so much to provide the equipment, but often, covertly, to help children to produce the right result by suggesting answers, offering them tools and equipment. It is the same as in the universities, if I go round many Engineering Departments I see students conducting experimental projects, with the aid of technicians. They are the same projects that are required every year, and the 'correct' results are known to the technicians and the students before they begin. There is no element of creativity or of imagination which is, after all, what we claim technology is supposed to be about.

In university, just as in school, the laboratory technician's job is all too often to suggest to the student, 'you do not want to do it like that, you will not get the right result', and the right result is the same year after year. It is not an exciting experience, it is simply getting by. And a result, the enthusiasm, the spark, the imagination, which many students bring with them because of their school experience, gradually gets lost. And if those students become teachers they are conditioned to repeat the same kind of experience.

Of course, if we are not obsessed with the right answers, we will not obtain the conventional results and we have to adjust the process of assessment so that we can recognize the new, different things we are getting instead.

The way that technology is portrayed on television with programmes like Young Inventors, Tomorrow's World (all countries have these types of series) attempt to do something about it. Young Scientist Clubs, electronics magazines, hi-fi journals, are all talking a language that young people can respond to and commonly lead to high standards not only of creativity but also of skill and 'finish'. It is not a simplified, easy language, it is sophisticated, but young people can use it and do respond. We do not always spark that sort of enthusiasm in the schools or in universities. I know it is difficult for schools to match media resources – but we should try to make positive links with them.

In the technology area, there is a particular problem. Many people feel sad that the conventional levels of skilled performances are sometimes not as high as they were, because students are not focusing as zealously as they once did on absolute precision and quality of finish. One cannot have it all. But certainly the goals we have been going for in the past have not always served us as well as they might have done. The essential feature is for students to be able to feel proud of their product and if skill is necessary, we must help them to acquire it.

3 The Role of Industry

Now let me turn to the second of my headings. We do not always require industry to share as fully in the process of technology education as fully as we should. This is despite many attempts to engage schools and industry more closely as described by Innes (1995), in this volume, 'There is no doubt if one want to show students the excitement, interest, attractiveness, and the economic rewards of being a technologist, then one has to find a way to introduce them to people who are actually working as technologists. We have to get far more of those people into the schools and help them to relate closely so that children will understand what technology is all about. This includes not just the excitement but also the routine work so they will really feel that they can understand the whole picture. We cannot eliminate boredom from industry any more than from schools. But we must justify it, not impose it.

Benson (1995), has described the key work in economics and industrial understanding at the University of Central England. At Warwick University too we have a large Centre for Education and Industry. We are running, with a number of other bodies, a whole series of schemes where we are getting industrialists into schools. In return we are getting teachers into industry, not just in some 'observing' role but to actually undertake projects in technology. We are also developing 'compact' arrangements, whereby children do not just go and spend time in industry, but actually get involved in an on-going relationship with a local industry, which guarantees employment, if they achieve specified results in accreditation. It is very easy to say that there are unfilled needs for more technologists but actually turning those needs into jobs which students can obtain is not easy as it seems. There has to be a real prospect of employment and it only becomes real when there is an understanding and engagement between industry and students, when each side knows what is really on offer. It is pointless hyping young people up for technology if we do nothing to ensure a reasonable level of certainty, of employment, a reasonably level of financial reward and status. All too commonly we have a situation of highly qualified, unemployed school and university leavers, and yet plenty of vacancies in industry. There is no point whatever in developing better technology education, unless we do something about linking it with the career, employment and labour market structure. It is a huge effort, but it is absolutely vital. And of course we need to realize that not all industrialists are unequivocally enthusiastic about young workers who know too much about how enterprises work, how wealth is created and profits distributed. We must also teach diplomacy and sensitivity.

4 The Role of Student

Now I am going onto my third area, and that is the young people themselves. Not because they are the least important, but because they are the most important. I am always uncertain whether to talk about pupils, students or young people. If I use all those terms, it indicates nothing more than my uncertainty rather than my confusion.

Somehow, one has to make technology related to their lives. This must include delivering not only jobs but feasible jobs, rewarding jobs, worth while jobs and status giving jobs. It must also be related to their lives as consumers, parents and citizens. Putting it simply, technology has to be perceived by young people as life enhancing and only then will they have the motivation and enthusiasm to learn effectively. In our best schools and colleges we know that technology is motivating young people, not just through self interest in being more effective consumers, having more interesting hobbies, or having a chance of a better job, but as a means of actually helping other people. For example there is huge enthusiasm among young people on technology courses for developing all kinds of strategies, aids and devices, that will help handicapped people, young and old, that offer better equipment in their homes and hospitals. Similarly, the enthusiasm of young people for environmental preservation technology and green technology generally is widespread and contagious. Technology can empower them. Here and elsewhere the altruism of young people is a hugely motivating factor, and we can help them to be more effective in their caring through technology.

I must end with some comments about the role of girls, because often it seems easier to achieve the kind of things I have been saying with boys who work with motorbikes, high performance cars and hi-fi. I do not want to do anything to diminish the technological enthusiasm of young men in our society, but we are often less effective in involving women in technology. Yet, we have, with enthusiasm, the capability to do something about it. Research (Dale, 1972) shows that girls are usually much more involved and much more successful in technology in girls' schools rather than in mixed schools. There are many reasons, but one simple and obvious one is that they are often 'put down' in mixed schools by the boys, which see themselves as more likely to be successful, and who crowd out the girls in their bid for teacher attention.

I was at an excellent technology lesson recently in a comprehensive school in Manchester. But when the teacher put the apparatus out she said, 'When you are ready come forward and collect the apparatus, there is not quite enough for everybody, so those who are ready can use it first and those who are not quite ready will be able to use it next'. There was a rush for the apparatus. At the end, all the boys but only two of the girls had apparatus. This happens time and time again. It is hardly surprising that some girls find technology less interesting. We have to do something about that and as most education is now in mixed schools it is a major problem.

So they are fundamental problems in helping students, boys and girls, children of all abilities, to see themselves as being able to succeed in technology and of course, assessment procedures, guidance procedures, support systems and the way in which we organize classrooms are crucial in ensuring that that takes place. It is exactly the same at the universities; we have many women science students but they are mostly biologists. At graduation day every year there are two or three brave young women, who have succeeded in obtaining an engineering degree. Because they are so exceptional they get a special cheer as they come to receive their award, otherwise the ranks are solidly male. They get almost the same cheers as handicapped students when they come to the rostrum.

Let me reiterate, in conclusion that, I am impressed by the initiatives in many schools and colleges, but alas they are not enough. The only way in which we can really progress, is to engage technology in the education and self image and life style of every citizen, in their roles as workers and as consumers in a modern society. We have to make every citizen technologically capable and literate. It is just as vital as all the other basic life skills that are taken for granted.

In the twenty-first century those life skills have to be enhanced by what we, specialists in our subjects, can offer. Only then will any country be equipped to move forward as a fully developed twenty-first century society. If we fail, we imperil our societies. If we achieve, we can all go forward together. Let me finally remind you, lest you ever think otherwise, that there is no basic human deficiency. We are all of us, at our different levels, capable of achieving, remember the examples I gave you earlier. Human capability can deliver it as teachers we can interest, motivate and engage. Putting children on the leading edge of the learning curve is a vast responsibility, one that I take more seriously than anything else in my professional career. We have a million miles to go, but it is a journey that we can and must accomplish.

References

1. Benson, C.: Economic and Industrial Understanding as Part of Design and Technology Education in the Primary Curriculum. Paper presented at the conference "Technology Education, Innovation and Management", Banská Bystrica, 1994
2. Cotton, A.: The Introduction of Technology Laboratories into Schools in the Disadvantaged Countries of the Republic of South Africa. Paper presented at the conference "Technology Education, Innovation and Management", Banská Bystrica, 1994
3. Dale, R. R.: Education in Girls' Schools. London: Routledge 1992
4. Innes, S.: Working with Industry to Enhance Technology Education. Paper presented at the conference "Technology Education, Innovation and Management", Banská Bystrica, 1994
5. Kimbell, R.: Evaluating Children's Perception in Design and Technology. In: Smith, J. (ed.) Proceedings of the International Design and Technology Conference 1994. Loughborough: The University 1994
6. Shield, G.: Researching Technology Education: Some Insights. Paper presented at the conference "Technology Education, Innovation and Management", Banská Bystrica, 1994
7. Wragg, E. C., Bennett, N., Carre, C. G.: Primary Teachers and the National Curriculum. Research Papers in Education. 4 (3), pp. 17-45 (1989)

Economic and Industrial Understanding as Part of Design and Technology Education in the Primary Curriculum

Clare Benson
University of Central England
Faculty of Education
School of Mathematics, Science and Technology
Westbourne Road Edgbaston, Birmingham B15 3TN, United Kingdom

Abstract. Economic and industrial understanding is part of everyone's life, including that of young children. They are aware of their parents' work or unemployment, of the work of visitors to their schools, and of the economic exchanges that take place in their local shops. Children will become consumers, workers and producers and it is important therefore that the experiences that they have at school prepare them for their economic and working lives. The recent introduction of economic and industrial understanding as a cross curricular theme to be integrated into the whole primary curriculum, has raised the awareness of both teachers and pupils of many of the opportunities that exist to discuss and question issues that are raised by investigating 'the world of work'. However, all too often these opportunities are lost on young children. Technology-industry projects are often focused on older pupils but surely, if young children are denied these experiences, there will be little to build on and extend in later years. It is certainly not the case that primary children are 'too young' to find the experience of value. Therefore this paper will seek to:

- outline briefly the nature of EIU
- identify the importance of the inclusion of economic and industrial understanding in the primary curriculum (In the UK, primary aged children are 5-11 years.)
- examine the relationship between design and technology and economic and industrial understanding in the primary curriculum
- offer suggestions of different methods for including the development of such understanding in the primary curriculum for schools and for Primary Teacher Education programmes
- identify characteristics of successful educational institution-industry link projects.

Keywords. Economic and industrial understanding (EIU), design, industry link projects, primary curriculum, technology education

1 Introduction

Enterprise, mini-enterprise, industry and economic and industrial awareness are all terms that have been more widely used and understood in primary education since the Education Reform Act 1988 (ERA) and the introduction of the National Curriculum in 1989 in the UK. Before this, the development of economic and industrial understanding with primary pupils was largely dependent on the curricula set by individual schools and teachers. Despite such valuable initiatives as Industry Year 1986 and 'teachers into industry' placement schemes, often links with local industry were founded through the personal friendships of an interested parent or teacher with a worker in a local firm. Indeed many primary teachers had never experienced any place of work other than his/her school and the thought of moving into a different, unknown environment was not one which was welcomed with enthusiasm. However, after 1989, it was clearly evident that if the aims defined in the National Curriculum were to be met, economic and industrial understanding would need to be included in the primary curriculum, not in an ad hoc way, but through careful and thorough planning to ensure that all aspects were covered. To support this aspect of the curriculum, the National Curriculum Council published further guidance in the booklet 'Education for Economic and Industrial Understanding (EIU) 4'. This guidance not only included a framework for implementation but also a series of case studies to show how different schools were integrating EIU into their curricula.

2 What is EIU?

Certainly, EIU is broader than 'making money' through mini-enterprise and it is important that a broad interpretation of EIU is promoted if children are to gain from its inclusion in the curriculum. When this theme was first identified in 1990, there was confusion as to the nature of its content. There was much to be covered with the new National Curriculum and teachers were having to clarify the nature of several new curricular areas. Moreover, some teachers thought that the main focus should be on mini enterprise and many schools started schemes such as growing and selling plants, creating and selling a school newspaper and cooking food to be sold in a school shop. In some cases an indepth study took place as to the real nature of mini enterprise, such as the skills that are needed and the organization of the production line, but often the pupils saw it just as a way to make money for the school funds.

What is it then that should be included in the curriculum to ensure that all pupils have access to EIU? In an Enterprise Awareness in Teacher Education publication (EATE) 1990, it was suggested that the following broad areas should be included:

economic understanding
: Pupils should develop an awareness of how resources are used, possible alternatives and how time, materials and skills can be allocated. Everyday, they are involved in making decisions about resources in the classroom in activities relating to, for example, science, art and design and technology.

industry
: Pupils should gain an awareness of the range of places where people work such as in manufacturing, transportation, distribution, retailing and service industries, domestic work and of issues relating to unemployment. In school, the children can study the range of jobs and places where people work (kitchen, office, classroom) before moving out into the community.

enterprise education
: This will involve the pupils both in learning about personal qualities associated with being enterprising and in taking part in small scale enterprises which should develop these qualities.

3 Why Develop EIU in Primary Schools?

The argument for the development of EIU with young children is strong. Throughout their lives, pupils will need to make choices and decisions relating to the economy as they become producers, consumers and citizens. It is important that these decisions and choices are based on accurate information and personal experiences. The children will be producers at work, they will decide how to organize their finances and how to spend their money and they will become informed citizens, making decisions about, for example, the effect on the environment of economic development. It is important therefore that the education which they receive helps them to make sense of their economic and, hopefully, their working lives. However, it is not only knowledge that is important. The skills and attitudes that can be developed through the development of EIU are those that are valuable across the curriculum and for the pupils' personal and social development. The pupils will be involved in practical activities, thus learning through personal experience. They will be involved in exploring values and beliefs and learn to make value judgements.

Some are against the inclusion of EIU in the curriculum for primary children, arguing that it has little relevance for them. However, before children go to school already they are, for example, participating in the economy and making decisions about how to spend their money. Anna Craft (1991) argues that if children are not given the opportunity to explore EIU and to understand different points of view, they are unlikely to develop an informed interpretation as to its nature. Fisher (1990) discusses the importance of the development of critical and creative thinking and problem solving skills with young children. Appropriate contexts for developing these skills is essential and design and technology, including economic and industrial awareness, provides one such context.

4 The Relationship Between Design and Technology and Economic and Industrial Awareness

With the introduction of the EIU theme, it was important that teachers realized that this was not a new subject to be taught, but a theme which should be identified across the existing curriculum. Curriculum co-ordinators were encouraged to carry out an audit in order to identify what aspects of EIU were being covered in each subject areas. Aspects not being covered could then be included in the most appropriate way.

4.1 Knowledge and Understanding

Certain key concepts can be identified, all of which can be developed through design and technology activities:

Co-operation	working to a common goal
Conflict	differences between people at work
Interdependence	the strength that comes from co-operation
Power	differences in power and authority
Change	from technology, social and economic causes
Division of labour	how tasks are divided between workers.

Further knowledge and understanding is identified in Curriculum Guidance 4 EIU (1990). For children aged 5-7 years it is considered important that they:

identify and make decisions about resources
   In design and technology activities, they explore and use a variety of materials and they choose which are appropriate for their needs.

understand some of the costs and benefits in situations relevant to themselves
   In design and technology activities they understand all too well that resources are limited in the classroom and that making a decision to do something may result in problems in other areas

understand that people have needs
   In design and technology activities the children discuss needs before suggesting solutions to a problem; they evaluate familiar artefacts; and they identify their own and others' preferences.

be aware that they are consumers and that this links them to people who produce goods and provide services

In design and technology activities, the children could be involved in surveys to find out people's preferences or advertise products or events in which they are involved.

know that people work in different kinds of workplaces, do different jobs and use different skills and understand how some things are produced, using different resources

The children can explore their school, a local shop or café as a workplace and create a workplace in their classroom as a design and technology activity.

understand how tools and technology contribute to pupils' lives at home and school

In all design and technology activities in the classroom, the children are involved in the careful use and choice of an appropriate range of tools. They learn to take care of them and to use them in a safe way. For pupils aged 7-11 years, this knowledge and understanding will be built upon and extended.

resources

Pupils will be encouraged to understand some of the implications of limited resources. In their design and technology activities they will learn to adapt, avoid wastage and look for appropriate alternatives when involved in making artefacts.

costs and benefits

In design and technology activities, the children will explore further the possible constraints and priorities when making decisions, the needs and values of those from other cultures and the effects of design and technology activity on for example, the environment.

the relationship between consumer and producer

In design and technology activities involving making products for others, for example, a book, a card cakes, the pupils will need to consider the needs and preferences of others and not to produce goods solely because they need and like them.

production of goods

In a design and technology activity, such as a mini enterprise initiative, the pupils will need to take account of costs including time, materials and use of equipment.

the workplace

In a design and technology activity involving the creation of a workplace such as a shop, a doctor's surgery or factory, the children may be involved in team work and will gain a better understanding of how the parts of a team fit together and the importance of all the parts of a team in creating an effective system.

4.2 Skills

In addition to knowledge and understanding there are certain skills which relate to EIU and will be developed through design and technology. These include the ability to:

- co-operate and work as part of a team
- lead and take the initiative
- communicate effectively
- collect, analyse and interpret data from, for example, a survey of favourite foods and use this information to produce an appropriate product for the 'market'
- make decisions about for example use of appropriate resources.

4.3 Attitudes

It is not only knowledge and skills that can, and should be, developed through design and technology activities The development of appropriate attitudes are important. These include:

- a concern for the effective use of scarce resources
- a sense of responsibility for the consequences of their own actions on the other members of a team
- a respect for others' opinions and decisions

5 How Can EIU Be Developed and Included in the Primary Curriculum?

It has already been stated that many teachers in primary schools in England and Wales felt overwhelmed by the content of the National Curriculum, and EIU seemed like another burden. However its inclusion need not involve more teaching time and the following process may be useful in indicating how this might be achieved.

A curriculum audit
Take the knowledge, skills and attitudes to be developed through EIU and match these against the curriculum. Where are the matches? What are the gaps?

The gaps
Can these be covered through any existing programmes of work? If not, can an existing programme be changed to cover the gaps? If not, what activities would be needed to cover these? Which year groups would they be taught in? When would they be taught? Throughout the process, it is important that all staff are involved in order that they have a clear understanding of EIU and its implementation throughout the school. It is not necessary for all staff to be involved in all

discussions but clear lines of communications need to be established to keep all staff up to date.

A policy

A written policy would then clearly identify how EIU is being developed in the school.

5.1 The Delivery of EIU

There are a variety of ways of providing for EIU:

through specific subject areas
> This is perhaps more usual in a junior school in England and Wales (children aged 7-11 years)

through cross curricular topics
> Many primary schools deliver parts of the curriculum through these and a range of titles, such as The local environment, Clothes, Shops, Celebrations, would allow for the inclusion of work on EIU.

through a specific focus on an economic theme
> This might be an appropriate way of ensuring gaps are covered. Starting points could include a visit to a place of work, a visitor from a place of work, a mini enterprise project.

Delivery of EIU can be achieved through activities spread over a period of time or through the allocation of a block of time such as a week.

However, it could be argued that the successful delivery of EIU is not just a matter of its inclusion in the curriculum for the children. Teachers need to have wide experiences relating to 'the world of work' in order that they can plan and deliver a more effective curriculum. The 'teacher in industry' placement schemes are one method of gaining these experiences. There are a variety of schemes.

Teachers can work in industry for a short period of time; they can work in industry on a project to link their school to their placement; they can work on educational material that industry can distribute to schools; and they can gain sponsorship to work in industry in vacation times to develop projects for school use.

It is possible, indeed desirable, that teachers not only have these opportunities once in the classroom, but during their training. Whilst many courses did include aspects of EIU in their courses before 1989, it became essential for them to do so after the introduction of the National Curriculum. One example can be drawn from the Faculty of Education, University of Central England (my own institution). All students on the one year Post Graduate Primary Certificate in Education (PGCE) course have two days in an industrial placement, develop resource material for use in school from the visit, and are aware of the coverage of all aspects of EIU in all subject areas throughout their course, via a focused permeation model. All students on the four year primary BA with Qualified Teacher Status experience activities related to EIU and are aware of how all aspects of EIU are covered in their course through clear documentation. In addition, the specialist Technology students spend a week in the third year of their course in an industrial placement.

From this, as an assignment, they develop a substantial resource pack, which they can use with primary children to develop specific aspects of EIU (Ager 1993).

6 Characteristics Associated with Successful Educational Institution – Industry Link Projects

If schools and teacher training institutions are to continue to develop successful projects with industry, what lessons can be learnt from studying successful case studies? Certainly some characteristics can be identified from such projects:

- an openness in both partners to come to a better understanding of each others' workplaces
- the development of good personal, working relationships between both parties
- a real partnership, where both parties feel that they have something to contribute and gain something in return. It can be that schools see industry only as providers of items that they cannot afford and industry sees it as 'having done their bit for the community'
- clear objectives for a link project so those involved are clear as to its nature and outcomes
- a willingness to allow links to develop over a period of time, not to expect to achieve miracles overnight. Everyone has busy schedules and making time for visits is difficult.
- the use of volunteer staff members in the first instance. It is unlikely that successful developments will take place if staff are forced into an initiative.
- a project which has developed from a real need, not just a 'one-off' experience to satisfy a short term curriculum need.

7 Conclusion

The introduction of a National Curriculum may have limited curriculum initiatives which broaden and extend the primary curriculum as teachers grappled with the delivery of vast quantities of content. Nevertheless, industry linked projects continued and pupils gained much from these. With the impending release of a new, 'slimmer' National Curriculum, it could be expected that teachers will, for example, have more time to devote to projects focusing on EIU. However, there has been a movement to go back to the basics, to English and Mathematics, and to concentrate less on the foundation subjects, including design and technology and themes, such as EIU. It remains to be seen what the future holds, but it would seem a retrograde step to ignore EIU, a theme which already has enriched the curriculum and offered opportunities to involve pupils in learning about the 'world of work'.

References

1. Ager, R.: Pupil material for Economic and Industrial Contexts. In: Primary DATA Journal. pp. 19-23 (1993)
2. Allen, M.: Let's talk shop. In: Child Education, Vol. 69 No. 11, (1992)
3. Craft, A.: Economic and Industrial Understanding: empowerment or indoctrination? Paper for Second Research Conference on EIU 1990, London PNL Press, 1991
4. DES: Technology in the National Curriculum, HMSO, 1990
5. Fisher, R.: Teaching children to think. Oxford, Blackwell, 1990
6. NCC: Curriculum Guidance 4 Education for Economic and Industrial Understanding. York, NCC, 1990
7. Ross, A., Hutchings, M.: Enterprise, Economic and Industrial Understanding Kit. London, PNL Press, 1990
8. Ross, A. et al.: The Primary Enterprise Pack. London, PNL Press, 1990
9. SCAA: Draft Proposals for Design and Technology. HMSO, 1994

Working with Industry to Enhance Technology Education

Sylvia Innes
The Standing Conference on Schools' Science and Technology (SCSST)
Professional Development Unit,
76 Portland Place, London W1N 4AA, United Kingdom

Abstract. The Standing Conference on Schools' Science and Technology (SCSST) is an independent national organization established to promote the development of Science and Technology education. Its network of some 50 centres, based throughout the UK, offer activities which aim to interest and inform young people about Science, Technology and Engineering so that they can see the relevance of these subjects to industry and everyday life. Through this it is hoped that more people will be encouraged to make careers as scientists, technologists and engineers.

Effective focused industry links play an important role in improving the educational experience for pupils, not least by enabling teachers to keep pace with scientific knowledge and technological innovation, which can relate to the body of knowledge being taught in the school curriculum.

These links also make a significant contribution to making more resources available to schools both directly and indirectly. From seeing the relevance and need, pupils are encouraged to raise their own personal skill levels.

Keywords. Curriculum development, school/industry links, SCSST, teacher training

1 The Standing Conference on Schools' Science and Technology (SCSST) and Science and Technology Regional Organization (SATRO)

For the past twenty one years, SCSST (The Standing Conference on Schools' Science and Technology) and its national network, SATRO (Science and Technology Regional Organization), have played a major part in supporting industry links by exciting young people about science and technology through the activities offered by the SATRO network.

For example, engineers with experience and credibility in the eyes of the young, working on real and relevant projects in the classroom, makes an important contribution to the technology education taking place. The presence of an outside facilitator enables a richer and more meaningful learning experience for all involved i.e., pupil, teacher, engineer. It has benefits:

- for pupils - by widening their experience and relating theory and essential knowledge to reality
- for teachers - by widening their own knowledge and skills
- and for engineers - by providing opportunities to inform pupils of the possibilities of engineering as a career.

Young Engineers is a national SCSST programme and is organized through school clubs for the 11 to 18 age range. The youngsters become actively involved in real projects which have practical applications and which are designed to meet an identified need. A significant number of these projects have resulted in products with the potential for commercial development. Students' achievements are recognized each year through National Awards Finals where their work is displayed and prizes presented.

1986 was designated Industry Year in the UK and this focused public attention on the value of industry education links. During that year, the award scheme CREST (Creativity in Science and Technology) was introduced. The scheme is designed to encourage the study of science, engineering and technology. It stimulates and supports industry linked project work, develops creativity, perseverance and application of knowledge to challenging problems.

CREST offers tangible recognition of achievement through an accreditation process which can be used by teachers to provide evidence of problem solving experience matching the requirements of the National Curriculum for Science and Technology in the UK. Over 70,000 pupils have benefited from the scheme so far and there is evidence of increased numbers each year. Individuals are motivated by active involvement and enjoyment of the challenge. The result is improved scientific and technological capabilities and skill levels. Awards are given on the results of successful projects which are:

- task orientated
- problem solving
- and require team work.

CREST contribution is in both academic and vocational areas and supports teachers and others in developing a richer experience for all involved. It offers opportunities for students to:

- solve problems from industrial and scientific contexts and apply knowledge and skills from theoretical studies to real situations
- gain first hand experience of high level work in industry
- develop key skills and attitudes valued and sought by employers
- develop confidence in dealing with professional personnel

- recognise the vital contribution of science and technology to modern society

The effect of Industry Year was to raise awareness about schools industry links and to encourage teachers to visit industry with their pupils. For many it led to good follow up activities in the classroom based on what they had learned from the visit. To gain most from industry links teachers needed some guidance.

"For world class technological skills we need world class technology teachers and teaching. This means a properly trained and well motivated teaching force. Industry is uniquely placed to help because it has the relevant knowledge, experience and expertise which schools can benefit from in order to provide a relevant, modern and challenging curriculum." [1]

'Learning from Industry' was a scheme devised by SCSST to improve the experience of industry/education liaison for both pupils and teachers. It was also seen as a means of producing new curriculum materials, related to business and industry, which could bring relevance to many of the topics taught in schools. Teachers were offered opportunities to spend a week or two in industry with the objectives to:

- look for industrial contexts with clear links to the curriculum
- help improve pupil's motivation by showing the relevance of what they learn
- provide opportunities for group work and problem-solving based on real life situations
- generate more effective links between industry and education by involving teachers and business people in purposeful activities

2 Curriculum Development

Curriculum materials based on process control in industry was an example of the stimulating and high quality work which could result from the 'Learning from Industry' model. As a relatively new aspect of the school curriculum, useful and interesting teaching resource were difficult to find. Even more difficult to obtain were resources suitable for all levels of ability and meaningfully related to local industry. Through this scheme, stimulating curriculum material was developed, which provided pupils with pleasure and excitement whilst learning the basics of Computer Control Technology and heightened their awareness of its place in industry.

The benefits for all concerned in this scheme were immense. Teachers returned to the classroom refreshed and stimulated by the experience. Pupils benefited from relevant new and interesting activities which the teachers were now able to provide. Industrialists were given the opportunity to have real involvement in school curriculum matters and, in many cases, liaising with schools was seen as career development for young managers.

school curriculum matters and, in many cases, liaising with schools was seen as career development for young managers.

Following the introduction of the 1988 Educational Reform Act in the UK and the National Curriculum, the Department for Trade and Industry and the Department for Education and Science, agreed a three year strategy to support the work related curriculum. SCSST responded by developing the national Technology in Context programme which was introduced in 1991.

In addition, the introduction of the Technology National Curriculum required schools to make major changes in syllabus content and delivery methods. Teachers needed effective help in a variety of ways to meet the challenges. Our research in schools during 1990 identified the following needs:

- new resource materials exploring business and industry contexts
- links with companies to provide 'real-world' contexts for activities
- help for teachers to plan and manage project work with a business dimension
- help with activities giving pupils first hand experience of business
- training for teachers in business concepts and skills
- accessible local support and advice.

There was a clear need to establish a national support service to help schools prepare for and teach the work related curriculum through the business and industry contexts in the Technology Curriculum and Economic and Industrial Understanding. The Technology in Context Programme addressed these needs by:

- developing resource materials for the work related curriculum
- providing local and national in-service training for teachers
- establishing a local advice and support service through the national network of around fifty SATROs (Science and Technology Regional Organizations)

3 Teachers' Training

The Technology in Context programme laid the foundations for the now established Professional Development Unit at SCSST. It aims to support and improve science and technology education in schools by providing in-service training for teachers and advice to business and industry on how they might assist teachers to meet the needs of pupils in ways which are relevant to the industrial world. The Professional Development Unit offers:

- In-service training for teachers which is designed to enhance and widen their knowledge and understanding of science and technology and keep them up to date with new developments.
- Guidance and support materials for in-service training
- Dissemination of industry related resource materials through in-service training workshops which relate the materials and activities to curriculum requirements

- Accreditation at post graduate level for teachers successfully completing follow-up work in the classroom
- Guidance to business and industry on educational policy and the most urgent needs for professional development support for teachers.

The provision of high quality and appropriate support is achieved through:

- identification of key issues which are seen as essential to the long term advancement of science and technology education
- establishment of the capability to design and develop national programmes to support the teaching of science and technology and to write materials associated with them
- development of effective strategies to disseminate resources and in delivery of in - service training on a national scale
- good working relationships with government departments concerned with science and technology education and industry links
- establishment of an accreditation system for training undertaken by teachers
- appointment of a high quality consultative group which includes members from industry, university departments, government departments and educational consultants of national repute in the field of technology education
- a monitoring and evaluation system established to provide evidence of achievements and quality.

3.1 Industrial Links

Currently we are working on a programme to develop Design and Make Assignments with and Industrial Focus for pupils of 14 - 16 years old both for the established curriculum and the new General National Vocational Qualifications which are being introduced.

The Identification of Design and Make Assignments as an important means of teaching Design and Technology is a useful starting point for a curriculum, which is both manageable and innovative. There is a particular need for us to develop these assignments with an industrial focus.

We plan to develop a suite of well designed and manageable tasks on which teachers can draw so that the curriculum can develop in a way which is attractive to pupils and allows for differentiation. Strategies for management and delivery of the assignments in the classroom will also be suggested.

The assignments will be developed for particular areas of the Design and Technology Curriculum:

1. Using resistant materials
2. Using mechanical control
3. Using electrical control
4. Using textiles
5. Using food

A framework will be used to produce assignments that are suitably demanding:

1. Knowledge of the problem
   - What is it and how do pupils get it?

2. Technical knowledge
   - including the development of fluency in drawing and modelling

3. Strategic skills
   - What are they and how will they be taught?

4. Tools, materials and equipment
   - What are they and how will they be taught?

5. Important values
   - What are they and how will they be taught?

The benefits of the programme will be:

For pupils the assignments will provide a key element in the new courses that will be developed from 1995. They will have opportunities to:

- experience links with industry
- be involved in real situations giving relevance to their learning by experiencing focussed industrial tasks
- work with personnel from industry
- have an insight into career opportunities in industry

For teachers the assignments will provide:

- a tried and tested way of teaching that meets the requirements of the curriculum
- resources and training
- access to accreditation on work done with children
- support from industry links

For industrial partners:

- high profile involvement in a national project
- focusing the minds of young people on industrial processes and practice by providing useful industrial contexts for Design and Technology
- up to date information and guidance documents on the curriculum
- the programme will present a positive image of industry...

The programme has industrial sponsorship from Unilever and the involvement of their operating companies throughout the UK is being arranged. The Department for Education will also add their support for the programme.

The assignments will be developed regionally using local companies and trailing will take place in selected University Departments of Technology and in schools.
A National conference is planned for October 1995 to display and describe the outcomes.

Finally, the materials will be published and disseminated through a national in-service training programme.

Working with industry does not always mean working in industrial contexts. We are currently working with companies whose interest is to ensure a well educated and informed work force for the future who are interested and equipped to take up careers, with a science and technology relevance. Examples of this are:

4 Programmes

4.1 Ciba Teacher Development Programme

A programme to support the planning and teaching of science and technology in primary schools - funded by Ciba with support funding from the Office of Science and Technology. Over the next three years around 10,000 teachers will have benefited from this programme.

Training courses have been designed to address the most urgent and relevant needs of teachers which has been highlighted in recent research. They aim to:

- increase teachers confidence in teaching science and technology in the primary classroom
- develop teachers' understanding of scientific concepts
- explore technological activities through practical workshops
- enable teachers to plan coherent learning programmes to meet curricular requirements

4.2 Continuity and Progression in Design and Technology from Primary to Secondary School

Ensuring continuity of pupils' technology education on transfer from primary to secondary school has been fraught with difficulty and it is clear that regression in achievement is not uncommon. The programme will address this problem in a three year programme which begins this autumn. The programme is sponsored by industry and government departments.

4.3 Number and Its Application

A programme which will begin later this year to support the needs of primary schools teachers with mathematics to support the teaching of science and technology.

Working collaboratively with industry is a central feature of the work of SCSST. Links with industry are not new, but are becoming increasingly important as part of schools' development plans. Effective links with industry can make a valuable contribution to the teaching of science and technology in schools and to understanding management strategies. Useful industrial contexts can make the teaching of science and technology real and relevant for pupils, leading to powerful motivation to learn. They can add focus to vocational studies and contribute substantially to economical and industrial understanding.

If children are to be well taught and motivated in science and technology, it is essential that their teachers are confident, interested and excited by what they teach in these very demanding subjects.

Equipping teachers to present this area of the curriculum in an exciting way is crucial to the motivation of students.

By working with industrial partners it has been noted that the esteem and motivation of teachers has been raised. They value the knowledge that others are keen and interested to support their work in the classroom.

Some companies have recognized that today's teachers hold the key to our prosperity in the next century, since it is they who must inspire inquisitive young minds to explore the scientific and technological world around them.

Through effective programmes of professional development which are supported by industry, teachers are able to build their professional skills and confidence to enhance their teaching of Science and Technology.

School and Economy - First Steps in Hungary

Sándor Kiss and István Matuz
Kölcsey Református Teachers Training College
Department of Technology
Péterfia 1-7, H-4026 Debrecen, Hungary

Abstract. School prepares for: information, communication, sciences and arts, life situations. Technology is one of the subjects preparing for life. Technology is not only creating and making but is also a way of learning how to economize with material, money, time, information and so on. Living in a society is living in a family and working in a place under the rules of economy.

Knowing the main rules of economy is a way to success in all fields. The main relations should be taught even in the elementary schools. The first steps in this direction are starting now in Hungary.

Keywords. Curriculum subjects, economy education, primary school

1 Background

In accordance with the social changes in Hungary changes in the educational system and mainly in its regulation are in progress. The earlier direct and simple system is dissolving. New types of schools and new curricula in them are going to appear. The spectra of the school subjects' activities are more and more colourful as well. New subjects nearer to the pupils' and parents' expectation arise. Some earlier subjects are modified or disappear.

In the above circumstances the necessity of a national core curriculum is a real need. This document is under preparation for years and has been reformulated several times. Now it seems to be getting its probably last form. Looking through the last versions it seems to be general, that the document cannot be based directly on the traditional way of thinking in subjects only, but the main fields of literacy/culture/ or of human knowledge and skills must be used. These fields can appear in the form of different subjects or moduls in the schools.

The above fields of culture can be classified in different ways. One point of view can be the historical background or treatment what is needed or not in order to teach the given field. The tasks arising from this approach were mentioned by the author on the EGTB Conference concerning the historical treatment of the subject technology (Innsbruck 1993). Another and more traditional classification can be due to the practical role of the fields in the every day activity.

The usually taught sciences and arts (mainly in upper primary and secondary schools) are really very important ones but are not widely used by everybody and everywhere. Part of them is used but other parts only by them who wants to be specialized in these certain fields. It is general that these subjects are in accordance with the historical differentiation of the sciences and arts (physics, chemistry, biology or history, literature, music and so on). The question concerning these subjects is: the separated or the integrated way of teaching is more useful.

Another group of subjects can be formed considering their basic role in communication or in information exchange. These subjects spread from the beginning of the lower primary up to the end of the general education. We start with reading, writing and counting which are the basics of the communication learnt at school. New horizons of communication are opened up by learning foreign languages. Also new ways of communication are given by learning of the elements of informatics which must be always widely concerned.

The third group of knowledge and skills can be formed by taking the needs of real life after the school into consideration. One of the main tasks of the school (beginning from the lower primary) is to prepare the pupils for the life practices. One of the subjects to be mentioned here is technology in wide interpretation. This field can have different names like technology or polytechnics, work, design, home economics and so on. They have the following in common: to create and make it, to investigate and choose, ... and to economize. Pupils have to learn to economize on material, on environment, on time, on information - all of these can be transformed for money. This means that the elements of economy must be introduced into the school curriculum right from the beginning. The subject technology is strongly bound to economy but other subjects like mathematics, history and so on can have connections as well. These are the usual subjects linked with economy, but there are many hidden possibilities as well: school excursions, school garden or shop what can be in the frame of 4H movement for instance, and many other possibilities.

2 Survey

Starting from the point of the importance of economy education even in the primary schools we made a survey in the region of our town Debrecen. The survey was made among pupils (boys and girls) at the end of upper primary (age 13-14). They were not specially trained, and their parents have similar social distribution as it is usual in our region (workers, farmers or others living on one salary - only 15-20% of entrepreneurs). Two types of questionnaires were used: one concerning the economizing thinking of the family, another aiming at the general economical knowledge.

About half of the pupils have already delt with the economical problems of the family, but less than 40 percentage of them heard about it at school. Almost half of the pupils' families make home economics plans regularly, but more than 2/3 of the pupils think it to be necessary. Concerning their involvement into the home economical planning, only 20% of them answered that they are involved but 60% of them would need it.

Concerning the pupils' own economical tasks they answered that nearly 30% of them have such a task regularly and 50% not regularly but it is mainly (35%) shopping. It turned out that they have opportunity for handling money individually mainly during the school excursions or in camps (75%).

Concerning their own (pocket) money the pupils answered that 30% get it regularly and 20% if they ask for it. The usage of this money is controlled by 50% of the pupils' parents. It is characteristic that nearly all the pupils (90%) having regular pocket money want to save a certain amount of it.

Concerning the general economic circumstances nearly all of them (80%) mention almost all of the incomes of the family. It is a pity that about only 25% of this information is related to the school. Most of them are informed by various sources like the family or friends or TV, radio, newspapers and so on. Similar distribution of information sources was found asking for the taxes or different money redistributions.

3 One Proposal

Taking into consideration the draft results of the above survey our aim is to work out a serial of modules that can be used from the beginning of the primary schools. Now we could find some similar efforts in the upper primary schools. One of these was made by the Ecology Department of the Teachers Training College (for upper primary) in the town Eger. Now here is the summary of the workbook for age 13-14 made by them, emphasizing that the topic can be dealt in lower primary as well.

Now here is a programme based on the above mentioned and similar works. The first step could be a playful training about the pupils themselves: I am a pupil, if; a member of the family, if; a buyer, if; and so on, depending on a certain point of view. Pupils should take into consideration what is the field which they are good at or weak. If they consider whether the things are important for them, can be bought or not, they will realize an important point of distinction. The above and similar trainings are interesting for all of us but they can be enjoyable even for lower primary school children as well.

The next step on a way of 'economy education' in the primary school can be the realization of the importance of the 'market'. Even the young pupils can know what are the products in their surroundings which are 'home products' and which not. This recognition based on the young pupils reading ability leads to the recognition of the international exchange. But the way of exchange is not so complicated for the young children, as it is in real life. It can be a good question what are the goods which they usually exchange among them directly. Are these goods only goods or not? There are lots of things exchanged directly by the family members or by the friends which are not goods but services for example.

A step nearer to real life is realizing the role of the money, because a lot of things can be shown that cannot be exchanged directly. We can speak about the history of the money which can be demonstrated in lots of ways. The present and probably the future role of the money can be shown mentioning the wider distribution of the different cards for example. Concerning the money controlled market the question of prices is that which must be put. What does the price depend on? What is the inflation? How can the different national currencies be exchanged?

One more step can be the question of economizing on different things: materials, time information, environment - money again! The questions of economizing are the ways of the future of our pupils. These knowledge and skills must be incorporated into all school subjects where it is possible!

A definite field, where the above type of thinking is obvious, is home economy. The family is the smallest and definitely strongest unit of the world wide economy. We can ask even the young pupils about the main functions of the family (economical, taking care of the children and so on). We should have time for explaining the historical role of the family. The question of distribution of the activities in the family can be very interesting for the pupils: what is it what they like or dislike in homework.

The last part of this training could be the analysis of the connection between the individual (family) and society. What does it mean to be unemployed? How to become an enterpreneur? For example, 'I am a businessman' training questions:

- Excellent ideas how to make money?
- How to realize the ideas?
- What income could I have?
- How could I spend the money?

The end of the training provides examples of successful famous businessmen, who are well-known all over the world.

It is obvious that the field of economy has an increasing importance and it must be included in the school curriculum. The question is what the proper form can be: to treat it as a school subject or to put the idea of economy everywhere in the school curriculum. The separate subject can be useful at about the end of school at about age 13-14. But this subject has to have a sound basis in pupils' mind. For introduction even from the beginning of the primary school, the elaboration of various short teaching modules seems to be a useful way. This work is a call for such an activity in order to collect the experiences in this field.

The Objects and Tasks of Technology Education in Comprehensive Schools in the Czech Republic

Frantisek Mosna
Czech Association of Technology Education Teachers (SUTV)
c/o Charles University Prague
Pedagogical Faculty
M. D. Rettigové 4, CZ-11639 Prague, Czech Republic

Abstract. A pupil may acquire only basic understanding, skills and relation to technology within the frame of technology education at elementary school. This is concerning the universal bases which are useful for a wide spectrum of technique, technology and practical life.
  One of the tasks of comprehensive technology education is to get the pupils conversant so they are able to solve current technical problems and situations which they meet in their life.
  It is possible to divide the objects of comprehensive technology education into objects of knowledge, skills and abilities in realizing the known methods of activities, creative skills and abilities, relations and attitudes in technology.
  To reach the objects of comprehensive technology education there is a conclusive and integrated importance of the concept and content of technology education, conditions for realization and teaching subjects oriented to technology (compulsory and non formal) necessary.

Keywords. Technology education, comprehensive schools, system of education

The experiences derived from teaching subjects which are oriented to general technology at our elementary schools and from the comparable teaching subjects in foreign countries show that a pupil may acquire only basic understanding, skills and relation to technology within the frame of technology education at elementary school. This is concerning universal bases which are useful in a wide spectrum of technique, technology and practical life.
  From the aspect of modernization of the educational system for the 21st century we start with the principle that technology education must therefore respect the environment in which the pupil is growing up, his experiences, preceding knowledge and interests. It must lead to a deep understanding of technology, to

appropriate use of technology, to manage the problems connected with simple use of technology in current life situations. At the same time it must allow, as an application of the teaching content of natural science and social science subjects, the understanding of the natural science bases, social and other wider aspects of technology. The technology education forms an organic part of the educational system at the turn of the millennium which makes an independent development of personality possible. If today's life of humanity is not possible without technical progress and technology (a condition of permanent development of a society), the people must use it to their benefit without an unfavourable ecological impact. Due to this reason it is necessary to teach them. It would be very simple to expect that the children living in today's overtechnologized world shall naturally understand technique and shall orient themselves in it. The ignorance of an orientation in technique may cause economic and ecological damages.

The society is obliged to make it possible for its members to have a corresponding compulsory technology education in the course of comprehensive school years within the humanization of technology and to create presuppositions for technology understanding and therefore a better orientation in life.

One of the tasks of comprehensive technology education is to get the pupils conversant so they are able to solve current technical problems and situations which they meet in their life, because otherwise they shall get into a conflict with the world around them. 'technology illiteracy' of an individual is a certain demonstration of discrimination and impoverishment of his other knowledge. The technology education supports technical thinking development, develops creative potential of the pupils and supports and creates such intellectual abilities which may be used in other spheres of social life. Intensively developed creative thinking is formed by solving problems which are oriented to practical life. This is what enriches a person and gives him better presuppositions to apply in an economically developing society. It leads to creating elementary habits of work safety, of keeping hygienic regulations and regulations of culture of work, it develops motoric abilities and teaches basic manual skills which are necessary for life. Compulsory technology education contributes to creating and deepening moral character features of the pupils, it naturally develops aesthetic feeling. It creates in the pupils basic imagination of production activity, organization and control, and of economic methods of energy, material and raw material use. It acquaints the pupils with the basic transport systems, information technologies and agrotechnologies which a person uses in his life. It allows to take a look into the history of technique including its effect on the society, and even into the preassumed development in the sphere of technology. It also allows the pupils, and the school-leavers, to self-realize themselves and to effectively use their free time. That gives them an economic profit as well as social profit. It creates a base for further education.

When determining basic objects of comprehensive technology education and when drafting the subject of teaching in the Czech Republic, it is necessary to lay emphasis upon:

- Educational character of technology education within the frame of comprehensive education.
- High share of practical activities of the pupils with technique, deepened share of activities as making decisions, construction, production preparation and technique use, technique valuation etc.
- Orientation of teaching to technology use and orientation in situations for pupils and current situations of enterprises, housekeeping, free time and social life.
- Recording of a comprehensive cycle of creative use of technique and technology in teaching.
- The meaning of wider aspects of technology, particularly economic and ecological problems.

It is possible to divide the objects of comprehensive technology education into objects of the level of knowledge, skills and abilities in realizing the known methods of activities, creative skills and abilities, relations and attitudes in technology. A pupil (school-leaver) of comprehensive educational compulsory school knows at the level of his knowledge:

- The meaning of the term of technology, the analysis of technology, its importance in history and at the present time.
- The basic features of materials, their analysis. He is able to express their meaning in the technology production and use on the basis if his own experiences.
- The most often used and prospective materials which are used in technology.
- The bases of activity and technique project during the production and use of simple technical objects, the bases of orientation in technical graphics and technical standardization.
- The knowledge of technical materials worked at by simple equipment and current machines at the production of simple technical objects according to the technical documentation.
- The substance of activity (principle) and construction extended and prospective technology which is used in production, at home, in free time and in social life.
- The coherence of technology with nature and society, with technical, natural and social sciences, it understands the coherence of technology and economy, of the life environment protection, the dependence of technology upon energy and material sources, it orients on the possibilities of technology for the development of a personality, its creator and consumer even for the life in society, it correctly valuates the importance of technology in history, at the present time and in the future.
- Regulations of safe activity with technology in a necessary extent.
- The basic concepts particularly of wood working, machine, electrical industry and the characteristics of the most extended processes oriented to technology.
- The bases of agrotechnology and agroecology.

A pupil (school-leaver) of elementary school on the level of skills and realization of known methods of activity knows:

- How to explain the basic features of technical materials on the basis of their structure with the help of knowledge and other teaching subjects.
- How to judge the appropriateness of concrete cases of technical materials use according to their features.
- How to project the production of simple technical products and how to decide on a method of technology use at the solutions of current situations with the use of technical knowledge and knowledge of other subjects.
- How to work out the most used technical materials at the technical products production with the help of simple equipment or the simpliest machines.
- How to judge and how to use suitably (himself or in a co-operation) technical products from the sphere of industry, construction, agriculture, household, free time and social life which occur very often, on the basis of explicit explanation of their purpose, substance and construction. He knows how to perform their service, maintenance according to a manual in accordance with technical, natural and social regulations and aspects, he must respect the economy, energy and material saving, life environment protection, aesthetic and humanity aspects etc.
- To keep general and concrete regulations of activity with the used technology including the regulations of work safety.
- To adjoin the activities of general production technical character to the appropriate profession.

A pupil (school-leaver) of elementary school at the level of creative skills and abilities knows:

- How to select reasonably the optimum of the current materials which are accessible for production of the stated technical product.
- How to seek possibilities of technology use, independently and in a team co-operation project simple technical products according to determining their purpose, to use them for technical thinking and the imagination development.
- How to select and apply, from the view of technology, optimum methods of activity at the production of extended technical materials with the help of simple equipment.
- How to use current technical products from the sphere of production, household, free time and social life, in an optimum way.
- How to seek actively his place for the technique use at home, in free time, in society, and later on in production. How to aim purposefully for the selection of other study orientation or to enter into practice according to his own technical abilities.

A pupil (school-leaver) of elementary school at the level of relations and attitudes to technology:

- Explains positive relations and attitudes to technology and activity with technique, to its production and use.

- He is ready to develop activities oriented to technique during his work and during his free time and at home.
- He understands the aspects of technique and production and social possibilities of the society at the present time and in history.
- He realizes the coherences of activities with technology with the necessity of moral characteristic development of its user, as for example responsibility, perseverance, the ability of team co-operation, keeping the technical standards.
- He is ready for the part of economical consumer at his own level.
- He is convinced of the necessity of applying the aspects following from the wider aspects of technology during the activity with technique, particularly the aspects of economy, environment protection, work safety and unexceptionability, technique humanism. He also acknowledges the possibilities and necessity of technology for the development of an individual and the society.
- He recognizes his own possibilities during his further study, particularly oriented to technique, or when he is entering an occupation in the spheres of technology.

The stated objects limit a frame of technology education content and the developed sides of pupil's personality at the comprehensive compulsory school.

To reach the objects of comprehensive technology education there is a conclusive and integrated importance of the concept and content of technology education, conditions of realization and teaching subjects oriented to technology (compulsory and non formal) necessary.

Development and Implementation of a Model for Technology Education in South Africa - The ORT-STEP Experience

Eli Eisenberg
ORT-STEP Institute
Department of Technology Education
Eskom Conference Centre, Private Bag X 13, 1685 Halfway House, South Africa

Abstract. In May 1993 the ORT-STEP Institute opened its door and officially stepped into South Africa to implement and advance technology education. ORT-STEP 's philosophy of providing a meaningful education that will ultimately lead to employment and independence, is achieved through a balanced approach to technology education and a strong emphasis on the crucial role of teachers. ORT-STEP views teachers as the core factor in any educational system, they are the agents through which change can be achieved.
The ORT-STEP Institute has three main areas of activity:

1. The Teacher Education Program that trains and retrains teachers in Technology Education.
2. The ORT-TECHNOLOGY college, which is used as a live laboratory - a real live testing field - for teachers to build experience and confidence.
3. The Resource Centre which continually develops Technology oriented courseware, hardware, and software and actively helps teachers implement technology education in their own schools, through a three dimensional support system.

ORT-STEP incorporates the theory, implementation and evaluation of technology education in a holistic approach that is fast proving to be a lighthouse to technology educationalists in South Africa. The talk will elaborate on how ORT-STEP has met the challenges and overcome the difficulties and problems that stood in the way to success and on how ORT-STEP will make a significant impact on the majority of the South African population.

Keywords. ORT-STEP Institute, teaching methods, learning methods, teacher training, technology education

1 Introduction

One afternoon in December, Mr Rugare Gapare returned to his rural home and presented his family with a gift: It was a dazzling display of colourful flashing lights that could be used as a very attractive decoration for the family's Christmas tree. Much to his disappointment, the family's response was that it was incorrect of him to have spent so much of their hard earned money on such an extravagant product. It took some time for Rugare to convince his family that he himself had made the display with his own two hands. He showed them how he designed and made the product, he explained how it worked and added with pride that if anything went wrong he could repair it himself. "And I thought that only machines and Americans could make such things" his wife said. By the time he left the village, Rugare had many orders from friends and relatives. They all wanted to purchase their own flashing lights, as Rugare could make the product and sell it with a profit at a price that was still 50% cheaper than similar products on the market.

Rugare is a teacher, who had just completed the Flashing Lights project on the Higher Education Diploma course at the ORT-STEP (Science & Technology Education Project) Institute, in South Africa.

2 Why is Technology Education Critical to South Africa's Survival?

We live in a technological society where not many people and very few decision makers know about science and technology. This could have serious implications on our future.

Research shows that there is a close relationship between the proportion of technology education in the school curriculum and the industrial contribution to the Gross Domestic Product (GDP). This can be demonstrated in the histogram (Fig. 1) for the year 1990.

If industry is indeed the key factor in building the power and prosperity of developing nations, it is now time that we co-operated more closely with industry to find out what its current and projected future requirements will demand. One of the main reasons for formal technology education is to secure qualified and efficient people who can participate in the development of the country and improve their personal quality of life and the quality of their children's lives in the future (ORT-STEP, 1994). Dyrenfurth (1994) states that: "While no one seems exactly sure of what the future will be, we do seem to be on a sturdy footing with two fundamental tenets. These are, in fact, among the few high probability truths that exist about the future:

– The future will not be what it used to be, i.e., ongoing change will exist.
– Whatever futures will exist, it is clear that they will be more technological than the nearest futures, i.e., there will be no abatement in the rate of technology infusion into our lives".

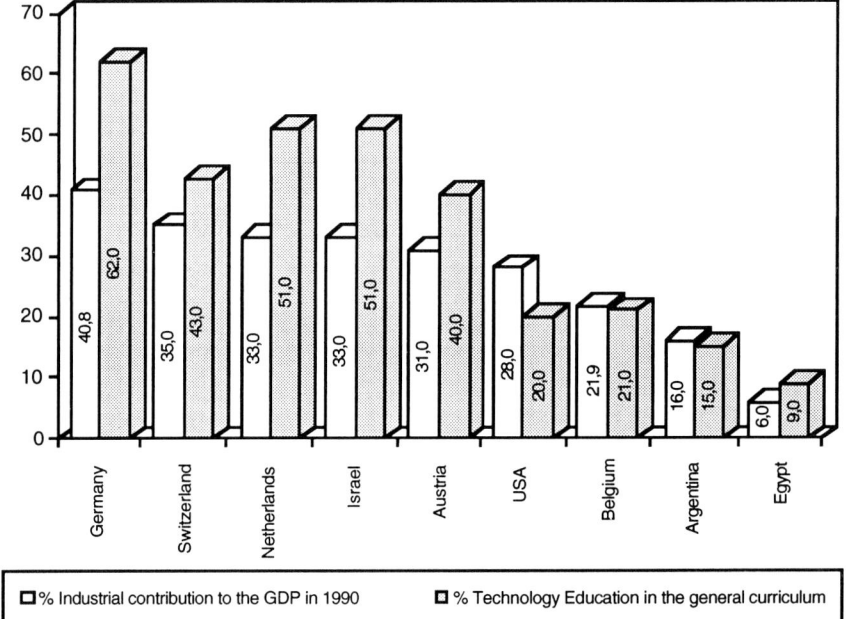

Fig. 1

Over 15 years ago, corporate America stopped asking itself how much money is wasted on staff training, but rather began asking how much money they were losing by not investing in training their staff.

Motorola is a good example of a company that has benefited tremendously from constantly investing a relatively high percentage of their gross turnover in the training, development and upgrading of its staff.

Is technology education and training necessary only for those involved in technological careers?

3 Technology Education for Whom?

The last decades of the twentieth century are characterized by the fact that the world in which we live has become more and more man-made. Man affects his environment for better or for worse, with the aid of technology.

Students of today tend to take the technological environment for granted. There is a tendency, even amongst adults, to accept many of the developments, inventions and technological creations as part of nature's laws.

Many are afraid to approach technology. Technology is frequently taken for granted. This results in an attitude where people are fearful of technology and are unaware that they can control technology or adapt it to suit their particular needs.

What indeed is the technological literacy required by people, be they doctors, lawyers, politicians, artisans and others?

What indeed is the technological literacy required by people, be they doctors, lawyers, politicians, artisans and others?

In modern technology the following skills are needed: cognitive ability, analytical thinking, problem solving and manual skills. To be technologically literate one should have a knowledge of technological skills and technological ways of thinking (Eisenberg, 1994).

Technology education world-wide has undergone important changes in the following areas:

Social: The conventional differentiation between the subject matter taught to boys (woodwork, metalwork...) as opposed to that taught to girls (sewing, home economics...) is no longer practiced. Boys and girls follow the same syllabi.

Content: Technology education involves cognitive as well as manual skills. Integrating knowledge of mathematics, business economics the sciences is required.

Teaching and learning methods: The computer has acquired a central position as a learning and teaching aid. It is used for locating, processing and editing information and for the control of technological processes.

4 A Model for Technology Education in South Africa

Prior to the development of the model for technology education in South Africa, ORT investigated, over a period of two years, the technology education field in South Africa, through networking and consulting with governmental and non-governmental organizations. The following are some of the issues that were identified:

- Historically, technical education suffers from a very low status in the educational hierarchy.
- The lack of suitably qualified technical teachers and teacher training resources.
- While the industrial community has developed a much greater awareness and need for technological skills, the focus is still on artisan training.
- South African society is generally perceived as being 'anti-technology' (Cotton, 1994)

One of the survey's main conclusions was that the success of any educational process is largely determined by the qualifications, motivation, self-confidence and self-esteem of the educators, teachers and trainers. They are the agents that face the challenge of change. If we wish to educate mankind on technological ways of thinking, we must first begin with educating the educators. This has been experienced over the years by ORT internationally, particularly in developing countries such as: Peru, Chile, Zaire and the Philippines.

The ORT-STEP Institute South Africa, has therefore set its main aim on educating, training and retraining teachers in technology education so that they become highly skilled, motivated and knowledgeable.

To achieve its aim of providing Technology and Science Education programmes, the Institute has established three interdependent facilities:

– A fully-fledged teacher training facility in Technology, Science and Mathematics.
– A Technology and Science education based secondary school.
– An educational resource centre for the development of technology oriented courseware, hardware and software.

The uniqueness of ORT-STEP is the closed loop feedback among the three components of the institute. This way the ORT-STEP Institute can offer a holistic solution to the educational needs of South Africa. Let us take a closer look at each of the three components of the model:

4.1 The Fully-fledged Technology Teacher Training Facility

A two year (480 contact hours) Higher Education Diploma course in Technology, Science and Mathematics accredited by a South African Tertiary Institution and the World ORT Union Organization.

The Technology curriculum of the programme is project based and is designed to impart to the student, the tools for thinking, planning, developing and evaluating. These and other skills are seldom suggested by the study of other subjects. The modes for developing these skills include:

– *Solutions to everyday problems in the environment of the student*
  e.g., building a flood warning alarm system. The design and realization of such a personal product also builds the students cognitive and creative ability.
– *Locating technological needs and environmental awareness*
  This encourages perception, research and a social responsibility in problem identification and resolution.
– *Selecting optimal over ideal solutions*
  This calls for critical thinking and analysis of constraints e.g., budget, time, material and knowledge. The student must be able to substantiate his choice.
– *Implementing ergonomic considerations in planning and design*
  The student will have to plan and shape the product for human operation in line with its properties and constraints identified earlier.

In conclusion, the processes which result in a technological product are multi-disciplinary. They involve both practical and concrete thought (production and operation) as well as abstract and lateral thought (problem solving and design). They also incorporate academic knowledge linked to mathematics, science and different disciplines in the field of Technology.

The technology project is experiential. This renders it invaluable in providing the tools for thinking as well as development of the skills mentioned above.

4.2 The Technology and Science Education Based Secondary School

The ORT College of Technology, with the full capacity of 300 students, has been established as a model school for matriculating in technology education. The main purpose of setting up the college is to serve the teachers in the ORT-STEP Higher Education Diploma course and to allow them to implement Technology lessons in a real life situation rather than a simulated environment or through role play.

There were no predetermined criteria for the selection of students into the college and most are average and below average. It is with these students that we can best test the strength of the ORT learning and teaching methodology.

While the importance of technology is stressed, it should also be emphasised that it is not more or less important than any other subject, and in addition to other core curricular subjects it is compulsory for students to learn Art, Music and Drama. This will broaden the horizons of South African pupils, rather than limiting them to a specific vocation in life and perpetuating the poor image of technology in South Africa.

4.3 An Educational Resource Centre

The resource centre has two main functions:

a. The development of educational packages, hardware, software and courseware and the adaptation of international educational packages for the South African environment and context. Here again, ORT-STEP enjoys the unique advantage of having the teachers as a source for feedback and for the development of material. The resource centre has already begun to establish a South African library for technology educational packages.

b. A support and backup service to the technology teachers, while they are on the course and more importantly once they graduate from the course and begin to introduce technology in their communities, so that they will not have to cope on their own.

The support is threefold:

1. A 'hot line' between the teacher and experts in the resource centre. This telephone support is not limited to school hours only.
2. Qualified tutors from the resource centre pay routine visits to teachers in their schools offering hands on support during the lessons. The feedback to the teacher is confidential and does not involve the school principal.
3. A technical service team is on call to maintain and repair the technology hardware and software.

Teachers on the course have already started to implement Technology modules in their own schools with a high level of enthusiasm and success. Many teachers feel that the programme has improved their overall teaching ability and given them important skills and tools to motivate and interest their students.

Pupils at the ORT College of Technology have taken the initiative to return to their previous schools in order to tell others about their experiences and to encourage them to apply for a technology education.

The main concern of educationalists in South Africa today is how to reach the masses of teachers in order to train and educate them in a short time, cost effectively while still maintaining high standards?

5 Greater Impact on the Majority of the South African Population - Quality and Quantity

Choosing the teachers as the most important factor in the educational system is largely due to the exponential growth that could be achieved by training teachers who, in turn, will train hundreds of students.

In keeping with that line of thought, ORT-STEP investigated other ways in which it could create an impact on the majority of the population, without dropping the high standards it has set itself. The aim is to achieve the quantity without compromising on the quality.

Three areas of activity have been identified:

1. Training the Educators

By training the lecturers at the teacher training colleges and educational faculties of universities, we reach teachers at an early stage in their training. It is aimed to train those teachers with a Technology educational component and create an awareness of teaching Technology, Science and Mathematics as an option thus creating a ripple effect.

2. Distance learning through Interactive Television

This system utilises technology in conjunction with human interaction. This involves the transmission of image, voice and data from a central point where the teacher is located, and a multiple number of remote sites, where the students are.

The technology for interactive distance learning integrates a number of technologies, such as satellite communication, compressed digital video technology and new technologies in personal computing.

There are a number of advantages to this method of distance learning:

– It is fully interactive, with the human touch allowing for creative and personalized communication. Using satellite, there are basically no limits to the number of remote sites that can be reached depending on the subject of the information being transmitted and the level of interaction required by the presenter.

- The cost of operation is relatively low.
- Students communicate with the teacher in the studio by using ordinary telephone units that are familiar and user friendly. They are linked to the system, independently of the national telephone network.
- Students are able to anonymously indicate to the teacher that they do not understand what is being said by simply pressing a digit on the telephone key pad.

The teacher receives a graphic presentation of the total percentage of students who do not understand the lesson at any point in time, and reacts accordingly.

3. ORT-STEP Science and Technology laboratories

By setting up technology and science laboratories in schools, it will be possible to fulfil the diverse needs of the community. During the mornings, the laboratories will be used by pupils of surrounding schools to study technology and science. In the afternoon they will serve as teachers training centres and in the evenings for Adult Basic Education programmes and other outreach projects for the communities. While Technology is more expensive than other subjects, utilising such centres extensively will make it cost effective and beneficial to the development of the communities in need.

6 Conclusion

It is clear that technology education in South Africa needs to be planned and developed to suit the unique requirements of South Africa. A national co-ordinating committee/ professional association for technology education should be established with the co-operation of all the relevant role players. Technology education and training should be in line with the Reconstruction and Development Programme (RDP) of the new South African government. This should be done on a national level yet should also meet the specific needs of each province. In terms of curriculum it is recommended that:

- technology education should provide equal opportunities for each individual student to develop his or her potential to the full.
- technology is not taught as part of science and that it is not considered as technical or vocational/professional training.
- technology should be taught as a subject in its own right with its own structure of knowledge which comprises knowledge items, skill items and ways of thinking.
- pupils learn to think in a more practical way related to real life situations and the workplace.

– most projects are graded and not structured tightly by the teacher. Each project should include room for the student to express his or her own identity and creativity.
– project work includes both investigation and innovation of technology.

Federico Mayer, Director-general of UNESCO, said "Human resource development is at the very heart of the developing process. Every person in society should have the advantages of a basic education which includes scientific and technological literacy" (Bowyer, 1990).

Rugare is only one example of how an educational process gained through technology education provides an individual with the knowledge and confidence to cope within a technological environment. The aim of technology education is to provide a meaningful education so that whether a school graduate chooses to go into industry, establish his/her own small business or continue studying at a tertiary level, he or she will be well equipped with the necessary skills. These skills, knowledge and technological ways of thinking will enable a school graduate to find employment, earn an income, and ultimately live an independent life, rather than depend on government structures for support.

References

1. The ORT-STEP Institute: International Survey on Technology Education. A survey report. ORT-STEP Institute, Halfway House 1994
2. Dyrenfurth, M. J.: The Impact of Technology Education on Economic Growth in South Africa. Paper presented at PROTEC, Johannesburg 1994
3. Eisenberg, E.: Technology Education and its Relationship to Science and Mathematics Education. Paper presented at the SAARMSE Conference, Durban 1994
4. Bowyer, J.: Scientific and Technological Literacy: Education for Change. UNESCO, Paris 1990

Present Days Problems of Technology Education in Slovakia

Ján Hudec and Ondrej Nemcok
Matej Bel University
Department of Technology Education
Tajovského ul. c. 40, SK-97401 Banská Bystrica, Slovak Republic

Abstract. In this article we briefly mention the development of technology education in former Czechoslovakia, we frame the aims of this education and we name both positive and negative features of Technical Education in the school system of socialism.
Then we deal with the consequences of the Velvet Revolution as far as the public opinion towards technology education and its realization in the process of general education is concerned. We speak of current problems of technology education in the Slovak Republic and indicate the direction of its future development. We try to suggest and defend with arguments the steps we plan to take in order to preserve and support this subject. In the end we express our belief that fruitful international contacts are inevitable in the field of technology education.

Keywords. Technology education, technological literacy, technical thinking, reforms in education

1 History

Technology education started to be taught at our elementary schools since the year 1960, first under the name Technology Activity, then retitled Working Education, the newer title being historically related to the former subject Manual Works which had existed in our education system since the beginning of this century. In the first years of the socialism in former Czechoslovakia when the working class was officially proclaimed the ruling class, the subject Technology Activity was established mainly to develop manual skills of elementary school pupils, in order to prepare them for Workmen's professions.
A considerably smaller attention was paid to their creativity, personal ability to solve problems successfully after formulating them once. These talents were being

grown rather in the sphere of so-called out-of-school activities practiced by official youth and children organizations, such as the House of Pioneers or Zväzarm, both subsidized by the state. At grammar schools technology education was taught as a part of a subject called Polytechnical Education, later retitled Introduction to Manufacture and Professional Training.

What may be taken as a positive of this kind of technology education is achieving a certain degree of manual skill with every elementary school pupil (girls including). However, the pupils did not acquire any technical thinking. They only imitated the teacher or instructor without any need to think, gaining their skill by purely automatic means. Manual skill is highly beneficial in a society with insufficiently developed public services (so-called 'do-it-yourself skill'), and last but not least it produces an almost universal readiness of the pupils f various occupations requiring this type of aptitude. There is also an impoi it feature not to be neglected: even today technology education offers the very needed relaxation from theoretical subjects and an important self-confidence, appreciation of one's own abilities.

What on the other hand represents the negatives of technology education as it used to be taught at elementary schools is the overwhelming uniformity of education not taking into account the individuality of any pupil and the totally collective teaching methods which both led to the final low standard of personal creativity. The whole process was utterly collective, i.e., each pupil manufactured the same product, the same technology was prescribed for everyone and so was the rate. The teacher practically instructed and demonstrated every single operation, telling the pupils to imitate him/her. Of course, some did it better, some worse, thus the final quality of the manufactured product varied. On this basis of the quality the pupils were marked. In order to make the education even more uniform, central projects of recommended products were issued, technologies were prescribed and in the centres of materials semiproducts were being prepared.

In fact the whole organization functioned more or less well and it covered every part of the country. After all that was the purpose of the unified school system which had been established in the very beginning of socialism. And not only in former Czechoslovakia. In former German Democratic Republic pupils in polytechnical centres even performed certain operations like hole drilling, edge grinding, bending, etc., on components sent them by factories. Many still keep remembering this practice with a certain nostalgy.

2 Present

After the Velvet Revolution, when the dogma of the ruling working class fell, Working Education and also technology education were automatically identified with the ideology of socialism. Even till nowadays we have not succeeded in convincing neither the public nor those responsible for this field of activity in the Ministry of Education, that technology education is a common general education subject in every well-developed industrial society and that it cannot be considered an 'invention of socialism'.

A team of authors [1] has proved this by the analysis of the state of technology education in the world. However, it seems that the results of their analysis did not appear in the public. And that is also one of the two main reasons why we have decided to organize this conference 'Technology education, Innovation and Management'. These are the reasons:

1. To prove the ability of our newly-founded Matej Bel University to join the family of universities all over the world
2. To show both professional and non-professional public in Slovakia that technology education is not an invention of socialism and that it is a common part of general education in the world.

Technological literacy - the main aim of technology education is necessary just as much as language, geographic or mathematical literacy. What might be a problem is the definition of the notion 'technological literacy'. It probably carries a different meaning in every education system. In Slovakia we find it necessary for every pupil (or student) to acquire gradually, from the simple to the complex, from the abstract to the concrete, at least two abilities:

1. The ability to understand the basic principles of technology gradually, according to the age and on the basis of:

- understanding of the principles of various technical toys
- understanding of basic principles of household appliances and their optimum use (considering the point of view of safety, ecology, power supplies, purpose...)
- understanding of various technical principles we meet with in our everyday life (at school, in the street, in the shop, etc.)

2. The ability to cope with everyday problems creatively, either individually or in a team, following with pattern:

- formulation of the problem (the heart of the matter)
- suggestion of possible solutions (always a variety of possibilities)
- choosing the optimal solution (optimalization)
- realization of the solution
- evaluation of the solution (analysis) and contingent corrections
  (if necessary starting from point a/ again)

And what do we imagine under the term technical thinking? For the time being, our subject bears the title of technology education. Its purpose is to educate pupils so that they would be able to think technically and behave technically in appropriate situations. If I were to explain it more thoroughly, I would use this example: many people, especially of the tender sex, have problems with the direction of turning of the tap if the water does not flow and the tap is partly open. As the tap has its limit positions on both sides, they do not know which way to turn it off. Turning the window handles could be similar. Finally the truth is found out, but only by trial end error, not purposefully. That is the way non-technically thinking people behave in daily life. Technical thinking means:

- to assume a technical, constructive attitude towards various problems not necessarily of a technical character in daily life,
- to have a positive relationship to the technology around us (at home, in the street), not to deny it a priori,
- not to see in the high technology something supernatural, beyond understanding, on the contrary, take it as a creation of human brain, a work of specialists,
- not to blame technology as the source of destruction of the environment, quite to opposite, by using technology appropriately we could save and improve the environment,
- to perceive technology as a part of mankind's life, designed to help the man, to make his life more pleasant and easier.

Technology involves all spheres of human life and activity. We should approach the problem of technology education from this aspect. The absolute necessity of technology education does not have to be stressed in this plenum and on this forum. That must be done in places where we are still misunderstood and where the structure and the content of our education system are being essentially influenced. However, if we want to persuade those institutions, we should not only issue proclamations but at the same time present concrete arguments and suggestions. If until now education has been directed towards manual activity, in the future it is expected to prepare the pupils for practical life besides. Its content shall orientate towards individually and socially important spheres of technology:

– Work and manufacture: technologies, forms of work, machinery, tools, materials forming of technical objects.
– Constructions and environment: construction works, bearing systems, projecting of constructions, ways of constructing, equipment of constructions, functional articulation of spaces.
– Supplies and pick-up: Energy-production and use, water supplies and draining off, production and recycling of goods.
– Transport: Lifting devices and vehicles, transportation equipment and conceptions of transport, safety of transport, roads, bridges and ports, etc.
– Information and communication: control systems, automatization of processes, information transmission, data-processing appliances.

We assume that technological literacy and technical thinking seen in this light could help the man considerably to take advantage of the intensely developing technology and not only in technical occupations. We dare to claim that especially in human-oriented occupations we meet with education to technological literacy only as pupils at elementary schools attending lessons of technology education. A negative attitude towards technology is not a good perspective to the future. It is beyond any doubt that only a catastrophe of The Earth could deprive mankind of technology but it is highly improbable that mankind would survive such a catastrophe. Therefore it is practical to learn to use the already existing technology and to create a technology more and more perfect and at the same time less and less harmful for men and nature. This process is impossible without technology education.

Technology education in Slovakia is under the pressure of current problems of the society undergoing a process of transformation. The problems are mostly these:

- extremism of opinion
- unfinished polarization of society
- economical problems of society

What we call extremism are the opinions which say that technology education is not efficient in a general education system at all. In Slovakia, the very fashionable wave of humanization of elementary school system is understood as a maximum development of humanities to the detriment of sciences and thus also technical subjects. The result is that technology education has been totally pushed out of the high school system and at elementary schools it has been reduced to a minimum level. There is a dangerous opinion saying that language literacy is inevitable for the generation of today (because we have to communicate with the world) but technological literacy is useless. Those who support this opinion should answer the question about what and through what shall we communicate? Just as a remark, language illiteracy, for example, is not lethal in contrast to technical illiteracy.

Unfinished polarization of the society and a seriously damaged value system have caused a situation in which incomes of physicians, technicians or teachers are deeply below the level of incomes of provincial politicians, bank clerks or lawyers. This influences the social position of these professions and obviously the interest of the young generation in choosing their carriers. It is clear that the Slovak society cannot live forever on international loans. We have to produce efficiently as could be possible only by means of new technologies. But such technologies can be managed only by sufficiently trained people.

Economical problems afflict mostly two resorts of the state: the resort of education and the health service. In the resort of education the lack of financial support already touches the basic needs of schools (energy, mortage, teaching aids). The development, production and service of teaching aids, all formerly taken care of by the state establishment called 'Teaching Aids' ended after the closing the firm. For the time being there is no alternative solution. Due to the insolvency of schools, the private sector shows no interest to get involved with such activities. While technology education is an especially demanding subject as far as teaching aids and material are concerned, we, the teachers, can only watch it die slowly. At the same time technology education should be, as a consequence of intensely developing technologies, one of the most dynamically proceeding subjects.

3 Future

It is our aim to bring our education system nearer to the European and the world one. The Slovak Society of the technology teachers is one of the founding members of WOCATE. We have built up an information system functioning among the teachers of technology education at elementary schools and universities. This system is based upon the voluntary of individuals working in hard financial and social conditions. It is namely they who need at least moral support in the sense that their effort is purely in accordance with the world trend. That is why we would gratefully accept professional help, represented for example by recommendations of particular commissions of UNESCO dealing with subject of technology education. We expect WOCATE to be the most efficient unifying article between members and UNESCO.

What is our idea of modern teaching of technology education? We presume the creation of concrete teaching topics on every degree of education. These topics would repeat spirally each time on a higher degree, more thoroughly and entirely.

For teaching technology education only such methods should be chosen that are based on creativity and practical realization of ideas.

First of all we have to realize that directive and collective style of teaching technology education is no longer possible. We have to make the subject that will educate children and youth towards technology, attractive not only for those who are especially skillful. To manage this, we have to let the pupils work with what they want to, while putting them on the right line.

It takes a system when pupils could choose the subject they want to manufacture. The teacher will instruct them, show them the way of working, help them to choose the right tools and so on.

It means that while performing a productive activity, the pupils have to be creative. A variety of teaching aids, brick-boxes, folders, etc. should be used to make education of various systems of technology more efficient and interesting.

It shall be an individual process of education, highly demanding for the teacher, who should both technically and pedagogically manage his task.

It is up to us, the teachers, to fulfil this task, to convince the public of the necessity of our subject. On this occasion let us learn from each other and interchange our opinions and experience so that the conclusion of our conference will lead to an essential change of the hitherto trends in school system and that technology education will become a firm part of general education.

One of our closest aims is to convince the whole society, slowly to arise the private sector and also the technically oriented public that it is necessary to start teaching technology education at elementary schools. In case we do not succeed we could expect to have enough interpreters, economists and lawyers, but we will lack people able to establish new technologies in industry, business, education or medical care. We are also convinced that the life of a technically illiterate person will be even more difficult after the year 2000 as it is today.

4 Conclusion

We shall gladly accept views and ideas concerning modern teaching of technology education of our foreign colleagues. We shall also gladly describe or demonstrate the organization of technology education here. We are seriously interested in joining bilateral co-operation or multilateral international projects oriented towards technology education teaching. We assume that our research and conclusions drawn out of it could considerably contribute to the content and methodology of technology education.

References

1. Mosna, F. et al.: Didactics of Technology Education. University Textbook. Charles University, Prague. pp. 9-18 (1992)

European Supporting Programmes - Experiences in the Field of Education and Training from the View of East German Universities and Enterprises

Frank March
COMETT UETP Thuringia
c/o Technical University of Ilmenau
Academic Foreign Office
Max-Planck-Ring 14/PE 327, D-98684 Ilmenau, Germany

Abstract. Experiences of a 1992 established EU-institution in the former Eastern Germany with the instruments of the European Community concerning education and training in the field of technology.

Keywords. Co-operation university-enterprise, technology transfer, COMETT, ECC

1 COMETT - A Supporting Programme of the European Community

What is COMETT?

COMETT stands for the European Community Programme on Co-operation between universities and industry regarding training in the field of technology. COMETT is a multi-annual programme of European Community financial support for initiatives designed to promote university-industry co-operation in the field of training for technology. It was introduced in 1986 and its first three-year operational phase was between 1987 and 1989. The second operational phase of COMETT is for a five-year period beginning January 1, 1990.

1.1 The Objectives of COMETT & the Network in Europe

COMETT II aims at reinforcing training in technology (particulary advanced technology), the development of highly skilled human resources and the competitiveness of European industry. It is centred on the changing skill requirements of industry and its personnel requirements which necessitate complementary action both in the Member States and at Community level. It is concerned with needs which can and should be met through co-operation at European Community level.

The specific objectives of COMETT as stated in the Decision are:

– The contribution of technology training to economic and social development

'to improve the contribution of, in particular, advanced technology training at the various levels concerned and thus the contribution of training to the economic and social development of the Community';

– Joint university-industry training efforts

'to foster the joint development of training programmes and the exchange of experience, and also the optimum use of training resources at Community level, notably through the creation of transnational, sectoral and regional networks of, in particular, advanced technology training projects';

– Training needs of small and medium-sized firms

'to respond to the specific skill requirements of small and medium-sized businesses having regard to specified priority measures';

– Equal training opportunities for men and women

'to promote equal opportunities for men and women in initial and continuing training in, in particular, advanced technology';

– Promoting the European dimension

'to give a European dimension to co-operation between universities and industry in initial and continuing training with regard to technologies, their applications and transfer'.

1.2 The Operational Components of COMETT

To meet the general objectives, COMETT focuses on *three* interrelated areas of action, each of which constitutes a *Strand* within the Programme as a whole. The Decision specifies:

Strand A: European network of university-enterprise training partnerships

'The development and reinforcement of university-enterprise training partnerships (UETPs) and the extension of the European network, both regional and sectoral, to further transnational co-operation, particulary in the following fields:

- in contributing to the identification of training needs in technology and resolving them in liaison with relevant bodies in this field;
- assisting and facilitating the development and exploitation of projects within the other strands of COMETT II;

- strengthening co-operation and inter-regional transfer between Member States in the development of initial and continuing training for the needs of technologies, their application and transfer;
- developing links in the form of transnational sectoral networks and bringing together projects from the various strands of the programme in the same area of training.'

Strand B: Transnational exchanges

Grants for transnational exchanges of:

- students undergoing periods of 3-12 months training in industry in another Member state. One of the important assessment criteria in the selection of projects submitted is the commitment of the sending university to the possibility of this training period in industry being recognized as an integral part of the student's course, taking account of the specific nature of national education systems and possibilities for such recognition under them;
- persons who have completed their initial training, either enrolled at a university or after graduation and as a transition between study and a first employment, taking up placements of 6 months to 2 years in a business undertaking in another Member State for the purpose of taking part in an industrial development project in that undertaking;
- personnel seconded from universities to industry and industry to universities respectively in another Member State to bring their skills to the industry or university in question for the training activities and the professional practices of the host organization'.

The duration of these secondments is between 2 and 12 months.

Strand C: Joint projects for continuing training in technology (particulary advanced technology) and for multimedia distance training

Three types of project fall under this heading:

- Support for crash training courses with a European dimension in technology (particulary advanced technology) designed for the rapid dissemination-by and in universities and by and in industry of research and development results in the field of new technologies and their applications as well as for the promotion, particulary for small and medium-sized enterprises of the transfer of technological innovation to sectors in which it was not previously applied.
- Support for work on devising, developing and testing at European level joint training projects in technology (particulary advanced technology) initiated jointly by different industries in association with the universities concerned in at least two different Member States in fields relating to the new technologies and their applications.
- Support for multilateral arrangements for training in technology (particulary advanced technology) initiated jointly by different industries in association with the universities concerned aimed at establishing systems for distance training

utilising the new training technologies and/or resulting in transferable training products.

Strand D: Complementary promotion and back-up measures

In addition there is a fourth Strand (Strand D), which comprises a range of complementary promotion, evalution and back-up measures. This includes support for preparatory activities, notably in the form of visits or meetings, their objective being the formulation of transnational projects.

1.3 Participation of the COMETT-Programme

COMETT is a programme for the support of training initiatives between universities and enterprises. For the purpose of COMETT, the definitions are listed below:

- The term university is used in its general sense to indicate all types of post-secondary education and training establishments which offer, within the framework of initial and/or continuing training, qualifications or diplomas of that level, whatever such establishments may be called in the Member States.
- The term industry (or enterprises) is used to indicate all types of economic activity, including not only large but also small-and medium-sized enterprises (SME), whatever their legal status or their manner of applying new technologies. The term also covers independent economic organizations, in particular chambers of commerce and industry and/or their equivalents, professional associations and organizations representing employers and employees.
- The definition of a small- or medium-sized enterprise (SME) is an enterprise employing not more than 500 people.
- The term 'training' is taken to comprise all forms of training at the 'post-secondary' level ('undergraduate', 'short-cycle higher education' etc.), and encompasses both initial and continuing training.

Participation is open to all these organizations from EC Member States, as well as countries belonging to EFTA (European Free Trade Association).

1.4 General Eligibility Criteria for the Selection of Projects

The following eligibility criteria determine the eligibility of projects for consideration for COMETT support:

University - industry co-operation. Any co-operation must involve both university and industry as defined in the previous paragraph.

A transnational European Community framework. Projects must be conducted within a transnational European Community framework, i.e., the project partners must come from at least two different Member States. The only partial exception to this is in Strand A, where the regional UETPs may be restricted in membership to one Member State only, but where the activities of the UETP must nevertheless be transnational in their objectives.

Post-secondary training level. Project must relate to post-secondary training, either at initial or continuing level. COMETT therefore covers both undergraduate and postgraduate training, training which may or may not lead to a formal qualification and continuing training of all types, including conversion training and distance learning.

Technology and technology-related training. COMETT projects must concern training in technology (in particular advanced technology), applications of technology or subjects relating to technological change and its implications. COMETT projects proposals which do not meet the basic criteria cannot be accepted. Given the competitive nature of the COMETT application process, meeting the basic criteria may not in itself be sufficient for acceptance.

1.5 Characteristics of Projects to Be Supported

While each Strand has its own specific set of characteristics, the following are the principal characteristics aimed for in COMETT projects:

Strong industry commitment and participation. The projects supported should be able to demonstrate that industry is firmly committed to their aims and outputs and that industry will participate actively in the achievement of those aims and outputs.

Synergy with other European Community and Member State strategies. The projects should complement (without duplicating particular actions) general policy strategies, with specific regard to:

- other European Community training programmes, such as EUROTECNET and ERASMUS;
- European Community Research and Development, where COMETT complements the actions undertaken within the Framework Programme
- European Community Regional Policy, where the relationship of the projects accepted with regard to regional policy initiatives will be of special importance;
- European Community Industrial Policy, particulary in regard to small and medium-sized firms;
- Member State policy in related fields.

Where applications for COMETT support relate to other European Community programmes, the application will be carefully scrutinized to ensure that any support granted for the COMETT project is meaningful and efficiently co-ordinated with the support granted by the Community under the other programme in question.

Balanced economic and social development. To conform with the desire for balanced economic development in Europe expressed in the COMETT Decision, the Commission will seek to ensure that the selection of COMETT projects takes full account of varying situations and needs within the Member States and EFTA countries. Due account will be taken of the size of enterprises, the extent of technological development in the country in question and the extent to which a project includes measures to surmount language barriers which might pose special obstacles to co-operation across national frontiers.

Bearing in mind these factors, one overall objective will be to ensure an efficient spread of COMETT support across enterprises and universities in all the Member States and EFTA countries.

Project content which meets the COMETT objectives. The projects supported should be capable of demonstrating that they can make a high-quality contribution to the achievement of the COMETT objectives. In examining projects, the Commission will examine the following aspects of content:

- Priority for new skills. Priority is given to training geared towards new skills, both in the growth technology industries and in the traditional sectors where new technologies should be applied. Technology transfer is also an area for particular attention.
- Novel and innovative quality. The projects should cover ground which - whether in terms of content, mechanisms, or links - is new, not only for the universities and the industry concerned, but for the countries concerned and for the Community as a whole.
- Interdisciplinary character. The Commission will take special interest in projects addressing new interdisciplinary skills required as a result of technological change, and in projects concerned with the training of trainers.
- Simulative and exemplary character. The projects should not only be of high intrinsic quality but should also be such as to simulate similar developments within the Community's industry and universities, and thereby serve as models for similar initiatives.

Projects should be designed in such a way as to be disseminated widely and effectively not only in the countries concerned but in the Community as a whole. Where appropriate, the commercial exploitation of the training products developed should be explicitly provided for.

Building of a Technical Educational System of a Non-University Type in Slovakia

Pavel Prokopovic
Technical University of Kosice
Faculty of Professional Technical Studies
Sturova 31, SK-08 001 Presov, Slovak Republic

Abstract. The trends in the development of higher technical education regarding economic and social changes in the countries of Eastern and Central Europe have been analysed [1]. The present contribution aims at showing on the example of Technical University in Kosice, Faculty of Professional Technical Studies in Presov in what way these problems are solved and what outcomes have been reached. It focuses on the rise, form, content, and aims of Faculties of Professional Technical Studies in Slovakia (especially on the Faculty of Professional Technical Study in Presov the first of its kind in Slovakia).

The aim of the paper is to inform about realization of the technical university studies diversification in Slovakia (or Eastern Europe as a whole) in accordance with the 1992 OECD recommendations. Different approaches to the origin of such faculties, legislative lagging behind (the old though still valid University Education Act does not comprise this type of study), problems with titles awarding and many others - these are the tasks to solve. Pointing out the origin of the Faculty of Professional Technical Studies in Presov the paper shows the solution of this problem in Slovakia. At the same time we want to initiate the public professional discussion at an international level on how to approach the establishment of professional universities in the countries of Eastern and Central Europe.

The ever increasing number of graduates from technical colleges (in Slovak: Priemyselná skola) continuing in their university studies has led to a shortage of experts with college and university education with practical orientation to help industrial manufacture. The graduates from technical colleges should have found their posts in practice, the sphere for which they had been trained. The shortage of experts with secondary and university education in technology is solved nowadays by establishing professional technical higher education establishments of technology - Faculties of Professional Technical Study.

In the sphere of technical education, there is an endeavour to introduce the following:

1. Bachelor study. It is also introduced from below by some technical colleges, as well as from the above by some universities.

2. Higher educational establishments of professional technical orientation are subject of a more detailed survey.

Paradoxically enough, the old Act on universities is still in force and does not know any such form of study. Experiments in the field may be helpful for those currently devising a new act at universities.

Keywords. Higher education, technical colleges

1 Introduction

The renaming of Institutions of Higher Technological Education to Technical University in Kosice in the previous years was not merely a formal act. On the one hand, it expressed the endeavour to codify high standards of the school and its competence in teaching and research. On the other hand, it took up an obligation to achieve European standards in structure and content of teaching. Moreover, gradually all forms of pre-gradual and post-gradual forms of study are introduced. Providing various requalifying courses for the unemployed especially in the sphere of computer technology, automation, regulation, doctoral study PhD, etc., all these measures testify to the fact that the university already complies with the above conditions. The humanization of study is going to establish the University of the Third Age, the City University, the Faculty of Economics.

Incubation centre, Cassovia Technopolis, with the Technical University and the Magistrate of the Kosice City is in full swing. The nearest aim is to constitute the Faculty of Physical Engineering.

Still the most news are not only within the Technical University in Kosice, but are within the whole system of university education in Slovakia, it is establishing the Faculty of Professional Technical Study in Presov.

The Faculty of Professional Technical Studies (FPTS) has been established in Presov as the first institution of its kind in Slovakia. The orientation for the time being is in mechanical engineering, prospectively electrical engineering is planned. Establishing such faculties is fully supported by the Government of the Slovak Republic, by the Ministry of Education, and the Ministry of Finance and will also be fully backed by legislation in the currently prepared amendment in the Act on Institutions of Higher Education which should become effective as of October 1994.

2 The Conception of the FPTS (Professional Universities) and Their Arrangement According to the New Educational Act

At the present time of the economical recession even the developed industrial countries are beginning to reduce funds for their institutions of higher learning which are searching for a pragmatic solution to adapt the educational system to the needs of the market mechanism.

The conception of the FPTS in Slovakia derives from the need to give the student not only the inevitable theoretical knowledge but also to lead the student to the direct application of theory in the system of university - student - practice. The regional character of studies and the connection with essential enterprise in the region are stressed.

An important part of education at the FPTS, which is different from the classical type of university education, is the connection with the manufacturing, agricultural, and social practice. This is manifested in two semesters of education in practical conditions that are provided without the seat of the school.

The first semester of practical classes, for instance, takes place in the third semester, i.e., in the beginning of the second year. In the future this may be shifted in the fourth or fifth semester. This would be a logical conclusion of the first part of study which is concluded by state final examinations.

The first semester of field practice takes place in the enterprise affiliated with the faculty, in a way similar to teaching hospitals. They are prominent enterprises in the region. With true conviction it may be said that this, in Slovakia and Czech Republic new form of university education, is going to be an inseparable part of the university system in these countries.

3 Faculties of Professional Technical Study and Their Aims

At present, the senates at technical universities may decide on establishing the faculties of professional studies. This happened in the Technical University in Bratislava which set up the Faculty of Professional Study (FPTS) in Trnava and at the Technical University in Kosice, having established the FPTS in Presov. The third one is the Faculty of Travel and Tourism at Matej Bel University in Banská Bystrica.

The first to emerge was the FPTS in Presov, but at the Slovak Technical University, even before that they enabled 150 students to start studying at the Institute of Professional Studies at the Faculty of Materials and Technology (MtF) in Trnava.

The founding universities of FPTSs suppose that the new Act on Universities will enable them to create Professional Universities out of the FPTSs in which the branches will be within the responsibilities of the faculties and the orientation within the departments. Professional universities could be regionally linked up with the institutes of technology in such a way that their immense advantages could be manifested for the following:

1. professional engineering with regional entrepreneuring intentions and erecting new small and medium enterprises by way of privatization and employing innovative technologies,
2. the support of two semesters of education in live laboratories of the practice,
3. big chances of most of the students to travel from their homes,
4. the possibility of exclusion of the FPTSs from the universities with a maximum making use of both part - timer and full - time teachers.

This gives the chance to make the university study accessible to a larger number of students with professional engineering for the practice in close co-operation in live laboratories of practice by acquiring financial sources from regions and through sponsoring of their enterprises for FPTSs. By doing this we try to achieve that the FPTSs are compatible with Europe and advantageous in all respects with the development of heterogeneous university education in Slovakia. For Slovakia, the contemporary number of top - quality universities is sufficient as it is in the centres of science, but all the more professionally erudite FPTSs are needed in industrial regions.

3.1 The Mission and Main Tasks of FPTSs

- FPTSs are a form of university study which is to complement the existing university system,
- is to create new possibilities of study,
- FPTSs are a result of the process of differentiation of existing universities according to the needs of the present entrepreneurial practice,
- to make up for the complex instruction of professionals for small and medium - sized enterprises,
- to make the time of study shorter, as compared with the university studies,
- to prepare the student to carry out his/her profession in a complex way.

3.2 Content and Forms of Education in the FPTSs

- a marked distinction of the study (as to form and content) in the FPTSs from university study (to exclude the duplicity),
- the content of study is practical oriented (semesters of field practice, the level of practical abilities to perform physical work of the graduates),
- the content of study is derived from the really indicated needs of entrepreneurial practice,
- the content and forms of study correspond with the interest of those who wish to study,
- the content of study is flexible enough (the choice of study branch).

3.3 The Content, Form, and Outcomes of the Research at the FPTSs

The position and aims of the FPTSs within the structure of universities give the research activities of teachers other dimensions:

- the tasks offered by the practice of enterprises,
- the character of tasks is given by the applied research, development, and application,
- commercial aims of the research are followed,
- the outcomes of the transfer of knowledge and technologies in the practice are observed,
- the transfer of the research outcomes and development in the teaching practice are observed,
- the share of students in the solution of admitted research tasks is higher than in TU,
- the degree of making use of the equipment in the FPTSs (laboratories, workshops, libraries, and the like) is higher.

4 Faculty of Professional Technical Study in Presov

The FPTS in Presov follows up the tradition of the Faculty of Mechanical Engineering Workplace in Presov which was established here in 1979 with two specialized departments in Presov (The Department of Instrument and Automation Technology and the Department of Robotics). In the period of twelve years a collective of capable teachers emerged here giving education to more than 800 engineers in the branches of automation technology, robotics, integrated technologies, nanotechnologies, and robotic technologies. A number of teachers come here from Kosice. Presov was the centre of robotization within the whole Czechoslovakia. The city hosts two considerably large Industrial Automation Works with a broad scale of manufacture of electrical and pneumatic automation elements, electrical machines, electrical drives, power electronics, rectifiers, etc.

Teaching is also done by experts from the Research Institute of Robotics VUKOV (in Slovak: Vyskumny ústav kovopriemyslu) and the International Business and Research Association ROBOT. We have here laboratories, operational premises, two new students hostels have been constructed housing for 900 - 1000 students. These were the main reasons for establishing FPTS in Presov where all conditions have been fulfilled in the sphere of teaching and professionalism, but also inevitable conditions for student life as to accommodation, catering, sporting, and culture. Local and district administration authorities in Presov are extremely helpful and may serve as a brilliant example of supporting universities.

The Faculty of Professional Technical Study will educate engineering in a four-year study. All of the subjects taught will be practice-oriented. The teaching here will have a different character compared to that in the Technical University. The attendance on all lectures and seminars will be compulsory, the weekly extent of teaching hours being 30. Examining will predominantly be carried out in written form. The aim of study will be defined by the final graduate profile: practical and even pragmatical knowledge and skills and capabilities. Graduates should predominantly work in the manufacturing sphere as operational engineers in production, projection, technologies, business, maintenance, quality control systems, environment protection, etc.

5 Conclusion

The FPTS is specific in that it educates professionals in mechanical engineering for the posts in technology and management in the sphere of enterprises. Their advantage is a thorough knowledge of problems existing in manufacturing enterprises and technologies, since under the guidance of tutors, the students spend two semesters in manufacturing enterprises. If we consider the great success of those faculties in Germany and the USA, but also in some other countries of Western Europe, then one cannot doubt that our graduates will find successful application in big, medium, and small enterprises.

The FPTS inaugurated in Presov on the 15th December, 1992 has one study branch so far - that of general engineering. Its orientation is closely connected with the manufacturing orientation and the requirements of joint-stock companies in the region, in which our graduates later will find workplaces. The spheres of competence between classical Technical Universities and FPTS will be divided as follows:

- Graduates from universities will be taught in close connection with science and research. Their future career will mainly be in research and development, academic institutions, state administration institutions, educational establishments, etc.
- Graduates from non-universities higher education schools will be practice oriented.

FPTS bring alternatives, distribution, graduation, regionality, and permeability within the 'unified' system of higher education. Industry and private entrepreneuring are in need of such graduates. The first graduates who will finish their studies in four years will show whether and how the above aims will have been fulfilled.

In connection with the already mentioned, we would like to inform participants of the Conference, take part in the discussion and to find out the reaction of the professional public on the following:

1. Elaborated new specific programmes and study plans for FPTS in Presov, which quite substantially differ from the programmes of the university studies of the respective subjects for the existing first two classes of FPTS in Presov,

and prepared programmes for the following two classes (the third and fourth, which have not started yet) at FPTS in Presov. We will give the detailed specification on the Conference. We will inform the participants of the conference about the details personally or on the Poster session.
2. Create a special form of co-operation with the industry. The students would go twice to factories for half a year during their studies. An enterprise would play the role of a "facultative plant" (in the similar way as a faculty hospital for medical students), where there is expensive one - purpose equipment at their disposition, together with professional workers, as well as external teachers of FPTS.
3. Methodological problems in teaching connected with a remarkable reduction of hours for lectures in favour of practical exercises.
4. Specific problems, connected with the creation, legislature, recognition of a title, finding its place in the High Education Law, of Faculties of Professional Studies (of professional high education schools of a technical type with several faculties at a Professional High Education School in the near future).

During the discussion at the conference, educational plans will be offered, programmes for individual subjects of the first faculty - FPTS in Slovakia, in Presov for a public opponent process, for pedagogues from the countries of Central and Eastern Europe, where such kinds of faculties and professional schools are being introduced, as well as for experts from Fachhochschulen in Germany and Polytechnics in Britain, where there is a lot of experience with this kind of technical high education schools.

We will also mention our experience with co-operation with other countries in connection with the building of FPTS in Presov:

– In the framework of the PHARE programme, where FPTS Presov, together with FPTS in Trnava and the already prepared FPTS in Banská Bystrica, (FPTSs being created 'from above' from technical universities) and 6 Secondary Professional Schools, which are applying for the creation of FPTSs 'from below') are methodically and materially supported in the years 1993 - 1995 in the framework of the project 'Development of non-university higher education'
- In the framework another PHARE programme 'Distance education network in Central-Eastern European Countries'
- In the framework of TEMPUS:

  1. 'Technology of teaching of non-university subjects and the usage of plastics technologies in the countries of Central and Western Europe' with England (Sunderland) and Germany
  2. 'Transducers Technology' with the universities in Belgium, the Netherlands and Poland
  3. 'Implementation of EC quality and environmental requirements in computer aided design, manufacture and transport logistics for higher educational engineering'
  4. 'Study of industrial relations'

- In the framework of COPERNICUS:

  1. 'New methods and instruments to monitor the power flow in power systems in non - sinusoidal and un-balanced state'
  2. 'Examination of functional qualities of grinding wheels'

- In the framework at the approved grant "Working out and improvement of curricula for Faculties of Professional Technical Studies"
- With individual technical universities (Fachhochschulen, Polytechnics) in abroad: Zwickau, Berlin, Kempten, Köln, Wuppertal, Sunderland, Portsmouth, Dublin, London
- The involvement of FPTS-s from the Czech Republic and other countries into above projects is an open question.

The problems of setting up FPTS, problems with their legislature, content of studies, determination of the profile of a graduate, recognition of a title and so on, are the current ones also for the Czech Republic, as well as for other countries of Central and Eastern Europe. Because these problems are topical ones, I propose:

- to plan a round-table discussion on the topic during Conference.
- At the same time I would like to inform you present of the preparation of the international conference on this topic organized by FPTS in Presov, in the framework of TEMPUS in Slovakia and the Ministry of Education Slovak Republic.

References

1. Jallade, J. P.: Undergraduate Higher Education in Europe: Towards a Comparative Perspective. European Institute of Education and Social Policy, Paris 1992
2. Lipski, T.: The Aims and Objectives of the Polish Association for Engineering Education. Australasian Journal of Engineering Education. Volume 2. Number 2, pp. 143-150, (1991)
3. Pivniak, G. G.: The Development of Creative Activity within the System of Technical Education. Proceedings from "East - West Congress on Engineering Education." Jagiellonian University, Crakow, Poland, pp. 448-450, (1991)
4. Dmochowski, Z.: Flexible Engineering Education. Proceedings from "East - West Congress on Engineering Education". Jagiellonian University, Crakow, Poland, pp. 348-351, (1991)
5. Minayev, A. A., Traube, E. S.: The Problem of Higher Engineering Education Improvement at the Doneck Polytechnical Institute. Proceedings from "East - West Congress on Engineering Education". Jagiellonian University, Crakow, Poland, pp. 194-196, (1991)

6. Gilicze, E., Michelberger, P., Tanczos, K.: The New Education Program for Transformation Engineers at the Faculty of Transportation Engineering of the Technical University of Budapest. Proceedings from "East - West Congress on Engineering Education". Jagiellonian University, Crakow, Poland, pp. 178-180, (1991)

7. Vishkov, Y. D.: The Development of Creative Abilities and Intensification of Motivation During Engineering Education. Proceedings from "East - West Congress on Engineering Education". Jagiellonian University, Crakow, Poland, pp. 77-79, (1991)

8. Pudlowski, Z. J., Rochford, K.: The Nature and Effectiveness of University Electrical Engineering Education. Australasian Journal of Engineering Education. Vol. 3 (3), pp. 45-61

9. Pudlowski, Z. J.: An Undergraduate Engineering Degree Program in Electromechanics. EEE Research and Development. No 1, pp. 1-18, (1992)

What Factors Will Influence the Wider Implementation of Technology Education in South African Schools?

Preliminary Evidence Arising out of the Kwazulu-Natal Pilot

Rodney F. Sherwood
Natal Education Department
Research and Development Unit
P/Bag 9044, 3200 Pietermaritzburg, Republic of South Africa

Abstract. This paper begins by describing the policy initiatives and other events which have led to the current interest in factors which might influence the possible implementation of technology education as a subject in South African schools. It then goes on to outline the nature of the Natal Education Department's pilot in Design and Technology and consider some of the issues that it seems to raise in respect of the wider implementation of technology education in South African schools. Finally it comments on some ways in which policy makers and planners might respond to these issues.

Keywords. Curriculum development, technology education, subject implementation, primary education, secondary education

1 Background

In November 1991 the Committee of Heads of Education Departments (CHED) in South Africa, published two policy documents on education for general comment. The second "A Curriculum Model for South Africa" proposed, among other things, that a subject called Technology be included as a compulsory component of the school curriculum in years 1 to 9 and as an option in years 10 to 12. The document set 1996 as the year of implementation despite the fact that the CHED, in its brief to the Core Syllabus Committee appointed to draft the guideline document for Technology, did not request the committee to investigate the implications of attempting to implement the subject within the prescribed time. This apparent neglect of the obvious problems associated with implementing a new subject like Technology on a national scale resulted in African National Congress (ANC) policy analysts and others questioning the proposal.

However, by November 1991, an exploratory initiative to investigate the introduction of Technology as an alternative for existing subjects in its "practical" curriculum in years 1 to 9 had already been launched by the Natal Education Department (NED) in KwaZulu–Natal. By the beginning of 1993, this had developed into a formal Design and Technology pilot. In early 1994 the Centre for Education Policy Development (CEPD) of the ANC was in the process of finalising its own policy proposals on education and sought to inform their views on the feasibility of implementing Technology nationally through reference to the NED's pilot work.

The formal pilot, launched by the NED in January 1993, bears many of the characteristics of that department and the schools it serves. For this reason the pilot's nature and context needs to be understood so that the issues raised by it can be meaningfully translated to the wider and often different settings presented by other schools in the rest of the country.

2 The Nature of the NED Pilot in Kwazulu–Natal

The NED pilot sample included:

3 Pre-primary schools
6 Primary schools (including Junior primary sections)
3 Co-educational secondary schools
1 Girls secondary school
1 Boys secondary school
1 Girls independent secondary school.

They all (except for the independent school) fall under the administration of the Ex–Department of Education and Culture: House of Assembly which, under the previous government, administered schools for white children. One school is an Independent School and the others are State Aided. They are all comparatively well resourced and are generally staffed by well trained teachers. The average class size is 30 - 35 students.

None of the teachers involved has received specific training in technology education and some primary teachers initially expressed genuine reservations about their capacity to teach technology education programmes.

The pilot was initiated as a response to a perceived need to review the "practical" curriculum of schools and not that of Science. In this regard, the curriculum in such schools makes provision for subjects like Craft, Handwork/Needlework and Basic Techniques in years 1 to 7 and Home Economics and Industrial Arts in years 8 and 9. The pilot therefore uses the time allocated to these subjects for Design and Technology.

Curriculum support for the pilot is supplied through the work of a Design and Technology Subject Committee while day to day management of the pilot is undertaken by a Subject Adviser. The Subject Committee decided to use the term

Design and Technology to describe the programmes developed through the pilot because of their concern to emphasise the place of thinking and design in the final material. However, Design and Technology is regarded as a form of technology education. Monitoring of the pilot is carried out through the NED's Research and Development Unit.

The NED exploratory work in Design and Technology which began in 1990 and which led to the formal development of the pilot in 1993 was initially informed and, to some extent, driven by:

- the 'new vocationalism' of the 1980's in the United Kingdom;
- the Royal Society of Arts' 'Education for Capability' manifesto; and
- the work of the Schools Council Technology Project in the United Kingdom.

However, since 1992, the pilot has been broadly influenced by developments in a range of other countries and, indeed, considers maximum exposure to international initiatives a prerequisite for developing a strongly indigenous form Technology education for South African schools.

Notwithstanding the above, the short term aim of the NED pilot is to obtain information which will assist the NED's programme of 'curriculum development" in Design and Technology and clarify issues related to the 'implementation' of the subject. It was on the basis of this aim that the following issues related to implementation were identified in the NED's monitoring report at the end of 1993.

3 Accommodating Technology Education in NED Primary Schools

Because of the way the NED has chosen to accommodate Design and Technology in the pilot schools there is a danger that it could be seen as Craft "dressed up in other clothes". It is true that Design and Technology in NED schools has, in 1993, reflected an Art/Craft bias and steps are currently being taken to strengthen the relationship between Design and Technology and other fields of study within schools - particularly Science, Mathematics and Computer Literacy.

However, the matter of accommodating technology education in schools needs to be seen as both a practical and philosophical one. In the first instance, technology education has to be introduced in ways that do not overload the existing curriculum and are relatively easy for teachers to adapt to. On the other hand, by associating technology education with a particular subject or collection of subjects one tends to suggest a particular philosophical understanding of what technology education is.

The NED views technology education as neither Technical Education nor Applied Science but, given the strong relationship it has with both these fields and the fact that the Craft/Handwork element of the curriculum is in need of review, the above approach to implementation seemed most appropriate. It also tends to be the approach taken in New Zealand, Australia and other countries.

In most other education departments in South Africa the degree of provision made for subjects like Craft, Handwork/Needlework and Basic Techniques in the first nine years of schooling varies. For example, the Department of Education and Training's schools allow for these subjects while departments in the Self Governing States do not seem to do so. The NED approach to accommodation might therefore not be possible in some other schools. However, the advantage in using such subjects to accommodate technology education in schools where they occur is that they can be replaced by technology education and the new subject is not perceived to be an "add on" to the existing curriculum.

The disadvantage of using these subjects is that they may result in technology education inheriting the low status and negative perceptions often associated with practical subjects.

A possible alternative is to use Science as a curriculum base for technology education although, from the perspective of the NED pilot it presents a similar set of problems. For example:

- Would such an arrangement require a greater allocation of timetable time? If so, where would it be taken from?
- Would it be possible to do justice to the demands of Science and technology education in such an arrangement? and
- What would become of Craft/Handwork subjects in such schools?

4 Lack of Understanding

The NED pilot suggests that among teachers, parents, students and school communities there is little common understanding of:

- what technology education is;
- how it might differ from existing subjects in the curriculum like Computer Studies, Physical Science and Technical Subjects; or
- what its educational purpose and potential might be.

Often technology education is linked in people's minds with skills training and motivated in terms of narrow economic and employment outcomes and this may adversely influence responses to the development of the subject in schools. Provision therefore needs to be made for the debate surrounding technology education to be pursued at all levels, including among school communities.

5 The Existing Curriculum Structure in Schools

Introducing technology education into the curriculum requires careful consideration of its implications for existing subject provision within schools. For example, if introduced as a completely new subject in the first seven years, it runs the risk of being seen as additional requirement within an already overcrowded curriculum. If introduced in the place of, or as an extra element of, an existing subject then one has to be very careful to assess the implications that such a decision would have on the subject concerned and the aims and nature of the technology education programme itself.

In years 8 to 12, technology education is likely to have to take on an increasing degree of specialization. What might these forms of specialization be and how would they be accommodated within the existing curriculum structure of Secondary Schools? In addition, curriculum change in respect of technology education needs to be seen by teachers and communities to:

- be educationally credible and justified;
- be necessary;
- be well planned and managed;
- accommodate existing priorities and needs; and
- adopt inclusive approaches to curriculum development and implementation.

6 Training

Lack of basic technological capability and professional training in this field is the most significant obstacle to implementation on any scale. In addition, acquiring an adequate understanding of and confidence in using technology, teachers also need to develop the ability to teach, plan and manage classroom activities and assess student progress in ways that facilitate effective learning through a variety of experiences and tasks.

It is highly unlikely that any existing cadre of teachers in South Africa could implement technology education without thorough training. In this regard thought should be given to:

- the need to train teacher trainers in this field;
- exploring ways in which teacher training institutions might be assisted in getting pre-service programmes off the ground;
- innovative ways in which existing teachers and schools might be given in-service and subject support in the development of Technology Education in their schools (i.e., the use of in-service vehicles to reach rural schools/setting up teacher and school support centres at colleges of education); and
- medium and longer term strategies for teacher training and development in this field. Here one should include support for the establishment of 'Centres for School Science, Mathematics and Technology' at key Universities and Technikons.

7 Cost

Although not as important as teacher training, the nature and extent of resourcing is an essential factor in determining the quality of eventual programmes. In most schools current resourcing is likely to be extremely poor with many having no electricity, being located far from business and industrial contacts, having no way of funding the recurrent costs of technology education programmes and no means of securing tools and equipment.

The following estimates of the cost of resourcing schools for technology education are intended to provide a rough guide and are based upon the following assumptions:

- that there are approximately 20 000 primary schools (7 400 000 primary students), 5 000 secondary schools (2 900 000 secondary students) and 100 colleges of education in South Africa;
- that approximately 90% of primary schools and 70% of secondary schools are without special facilities to teach either Science or Technology (i.e., a laboratory or general work room at secondary level);
- that around 90% of primary schools and 70% of secondary schools have no adequate equipment for teaching Science or Technology and no way of meeting the recurrent costs (materials, chemicals etc.) of teaching Science and Technology if they did offer these subjects.
- that in cases where resources were allocated to schools some provision would have to be made to ensure that stock can be secured.

Given these assumptions, the cost of funding resources for Technology in South African schools might be as follows:

7.1 Primary Schools

Given the following profile of an average South African primary school:

Size of school:	370 students (7,4 million/20 000)
Average class size:	40 students
No. of classes:	9

and assuming that the most economical way to resource primary schools would be to provide a Science and Technology trolley (valued at R 6 000) for every five classes in a school, then each primary school would require two trolleys at a cost of R 12 000.

Estimated cost of trolleys
(for 90% of schools)   R 216 million

7.2 Secondary Schools

Assuming that secondary schools would need at least a basic laboratory/work area for Science and/or Technology then:

120 m² facility x R 1 200 per m²;
plus furniture, tools and equipment approx. R 90 000 for each school.

Estimated cost of facilities and equipment (for 70% of schools)
R 819 million

7.3 Colleges of Education

Some Colleges will already have adequate resources for Science but very few will be equipped for Technology. Assuming that all are not adequately resourced and would need, at least, the sort of provision necessary to equip some student teachers for service at Junior Secondary level.

240 m² facility x R 1 200 per m²;
plus furniture, tools and equipment approx. R 170 000 for each college.

Estimated cost of facilities and equipment R 45,8 million

Total estimated cost of resourcing schools and colleges for Technology
R 1 080,8 million

This cost does not take into account the need to secure (i.e., through provision of storage space, burglar guards, cupboards, locks etc.) or insure the resources or equipment provided.

The following estimates on the cost of training teachers for Technology education are intended to provide a rough guide and are based upon a limited programme which would involve appointing two extra staff at each college of education for this particular purpose.

200 teacher trainers x R 90 000 each per annum.
R 18 million per annum

If each college was able to service 20 primary schools per year the programme would have to run for ten years before staff in the current 20 000 primary schools received some in-service training. No allowance is made for growth in school numbers or inflation.

8 Access and Equity

For technology education to transcend the limitations of narrow skills development programmes or the uncritical learning of theory it will need confident, flexible, well trained teachers and appropriate resources.

Teacher training will take time and implementation patterns will, to some extent, be determined by the availability of staff and support of management in schools. Similarly, the quality of programmes will be dependent on the extent to which the state is capable of meeting the kind of costs outlined above.

The rate of teacher training, patterns of teacher deployment and the extent of state funding for technology education will be factors affecting access to and equity of provision in this field.

9 Relevance and Local Needs

Rural and other disadvantaged communities will experience particular difficulties in sustaining rich and varied technology education programmes. Their isolation and relative poverty will limit opportunities to link school learning with experience of technology in the wider community. Special provision needs to be made to ensure that such communities are not adversely affected by such conditions.

Some may argue that the technology education curriculum for rural schools needs to be specially fashioned to enhance its relevance for such students. While a Technology curriculum must obviously extend and develop a student's understanding and appreciation of technology within the context of the school and local community, in South Africa we must however ensure that it also enables the student to transcend the limitations and constraints of a particular technological setting. This principle applies equally to all students whether urban or rural, advantaged or disadvantaged.

10 Gender

Evidence from the NED pilot, from national debate and international research suggest that gender issues do and will continue to influence the nature of Science and technology education and the degree to which girls and women have meaningful access to learning in these fields. Research into this and the issues of language and classroom management referred to below, should form an integral part of development programmes in this field.

11 Technology and Second Language Learners

There is some evidence that students learning through a second language experience difficulty understanding technological terms and concepts and are so disadvantaged. While no specific steps have yet been taken in this regard, the NED is planning to examine the position of second language learners as part of its monitoring of the Design and Technology pilot in future.

12 Practical Classroom Issues

The NED pilot identifies three particular classroom problems that are likely to influence the quality of technology education programmes in schools. The first is the impact of class size (around 30 students in NED schools and 48 as a national average) on the teacher's capacity to manage, supervise and assist the learning process and ensure safety within classrooms during technology education lessons. This situation is aggravated where equipment is in short supply and/or power tools and equipment are being used.

The second has to do with the cost and difficulty of acquiring regular supplies of consumable materials for project work. Schools are not allocated funds for this purpose and yet access to such supplies is vital for the effective planning and quality of project work. Many schools are not in a position to obtain such materials from local businesses and expecting students to supply the materials often creates a range of other problems.

Finally, wherever possible, schools need to link their work in technology education to the application of technology in the wider community. Individual teachers have tended to find that establishing such links is difficult. While this may be a result of inexperience it is also likely that local business, industry and service agencies are simply not accustomed to accommodating the kind of links envisaged. For this reason, it may be necessary for local education offices to initiate such links and facilitate the development of collaborative projects - at least to start off with.

13 Conclusion

The NED pilot has been designed to investigate issues related to the development and implementation of Technology as a discrete subject or field of study in schools – but this is not the only form of accommodation being mooted in South Africa. It has been argued that Technology should be integrated into a revised Science syllabus in years 1 to 9, while other educationalists propose that Technology be taught across the curriculum – particularly in the primary school years.

Given the limited pool of expertise in technology education, piloting is the only credible way of testing the merits of these and other alternative forms of accommodation in South African schools. Furthermore, piloting provides the most appropriate mechanism for drawing practitioners into the curriculum development process and combining sound curriculum development principles with staff development.

As regards implementation - given the preliminary issues raised in this paper - it seems clear that technology education could not be implemented on a national scale in the short term and, in this sense, criticism of the CHED's Curriculum Model for South Africa is justified. Evidence of the difficulties associated with the

United Kingdom's attempt at rapid implementation in the early 1990's provides ample warning of the dangers of forcing changes on teachers who are not fully equipped to cope with them. However, this is not to suggest that implementation is impossible. The implementation of Technology should, instead, form part of a deliberate longer term national goal. The framework for such curriculum development should be so structured that it sets meaningful time frames for the:

- drafting of syllabus documentation and teacher support materials;
- development of teacher in–service and pre–service programmes;
- establishment of research centres in selected Universities, Technikons and other institutes of Higher Education; and
- systematic 'stepping out' of provincial pilots based on research findings, development targets and teacher readiness.

A strategy of this kind might seem too gradualist for those who would prefer to see curriculum change linked to short term political objectives but, in our context, the educational advantages of pacing development according to the growth of teacher capacity on the one hand, and the perils of setting unreasonable targets for implementation on the other, suggest that it presents the only rational and constructive route forward.

The Introduction of Technology Laboratories into Schools in the Disadvantaged Communities of the Republic of South Africa

Arthur Cotton
ORT-STEP Institute
Schools Consultancy Service
Private Bag X 13, 1685 Halfway House, Republic of South Africa

Abstract. The paper begins with a brief historical overview of vocational and technical education in South Africa. It then goes on to discuss the current situation with regard to technology education. The Government of National Unity has made a firm commitment towards a more relevant education system and sees technology education as a key element in the proposed restructuring of the South African education system. There is a lack of expertise in technology education and a great deal of debate and uncertainty as to how technology education should best be introduced.

In order to support the teacher training programme the ORT-STEP Institute has established a Schools' Consultancy Service. One of the main functions of this service is to establish technology laboratories in selected satellite schools. The paper goes on to discuss the funding, establishment and subsequent support of these laboratories. Topics covered include the role of the school principal, staffing, the technology teacher, the role of the school community, activities to be covered, buildings, equipment, links with other subjects, outreach programmes and the use of external resources. In addition to being responsible for the satellite schools, the Schools' Consultancy Service assists any school wishing to introduce technology into its curriculum. This is done on a paid consultancy basis. The paper discusses the work done in this field.

Keywords. ORT-STEP Institute, ORT-STEP satellite schools, ORT STEP Schools Consultancy Service, technology education

1 Introduction

The South African education system is currently in crisis and there are many issues which need to be addressed. One of the areas to be addressed is the question of relevance in the school curriculum. The African National Congress and the Government of National Unity have both supported the need to introduce technology education in principle but are faced with a number of difficulties. Resources, both in human and financial terms, are limited. In addition, the government is in the process of establishing itself and is preoccupied with the need to restructure education and address the historical imbalances brought about by the apartheid era.

2 A Brief Historical Perspective

Technology education in South Africa suffers from a bad reputation. The reasons for this are many and varied. Many of the reasons have historical roots and lie in cultural attitudes.

As Malherbe [1] says "...the Dutch Reformed Church was the first to propose vocational education as a measure for combating Poor Whitism. In the 1890s it sponsored the introduction of industrial schools and extended them after the Anglo-Boer War as a means of training potential Poor White boys from the rural areas in industrial occupations such as shoe-making, carpentry, smithy work etc., and training girls in domestic work. By 1910 there were only 400 pupils all told in these schools. ... In 1911 the Prisons Department established two industrial schools, more or less as reformatories, for destitute and delinquent children. In 1917 the Union Education Department took over the administration of the Children's Protection Act and, with that, all these industrial schools. Later it started vocational, agricultural and housecraft schools of its own. But these schools had a very unhappy tradition to live down. The fact that vocational education had been associated with the destitute, the defective and the delinquent sorely handicapped its future development. This association and the idea ..... that manual work was '(inferior) work' placed training in occupations requiring manual skills beyond the pale for the boy and girl from the well-to-do and average homes. Luckily that stigma gradually wore off. But historically these efforts at vocational training were conceived in charity and with the idea of redemption. They were born out of poverty, depressions, wars and epidemics."

Malherbe goes on to observe that, "This charity tradition was doubly unfortunate in that (a) it was bad for education and prevented it from meeting the educational needs of adolescents, and (b) it was bad for industry. We could hardly expect our trades and industry to be efficient if we chose to recruit chiefly the weak and maimed into the vocational field and the strong and healthy (mentally and morally) into the field where learning is supposed to be pursued for learning's sake."

The administrative dualism whereby technical education at the secondary level was administered by the central government, whereas the more desirable academic, university-directed education was administered by the provincial education departments, did nothing to improve the situation.

Given the history of technical and vocational education in South Africa it is not surprising that many continue to turn away from such education and regard it as automatically inferior.

3 The Present Position with Regard to Technology Education

Whilst there is fairly widespread agreement about the need to introduce technology education to the primary and secondary school curricula, there is considerable debate over the nature of technology education and the manner in which it should be implemented.

There are those who feel that the available resources should be channelled into the restructuring of education. They argue that the present human, financial and physical resources do not permit the introduction of an additional subject, particularly an expensive subject such as Technology, into the school curriculum. They propose that Technology should be introduced through subjects which are currently in the curriculum and that, in particular, we should 'technologise science'. It is argued that this would be cost effective, readily implemented and have the added advantage of making science more relevant and accessible. Those who oppose this view feel that science teachers are not trained, and are often ill-suited, to introduce Technology. They would also argue that the current science curriculum needs to be upgraded rather than diluted by the introduction of a technology component.

On the other end of the spectrum there are those who feel that, whatever the present difficulties, the economic and educational needs of the country are such that we cannot afford not to introduce technology as a subject in its own right in the primary and secondary education of all pupils. They accept that this will have to be done incrementally. It is argued that Technology is a discipline with its own body of skills and knowledge and its own way of thinking. It is seen to have the advantage of being one of the few disciplines which links head, heart and hands in a purposeful manner. The proponents of this view argue that all pupils need to develop an understanding of the nature of technology in order to become effective citizens in the technological world in which they live. Furthermore, a holistic and well balanced technology curriculum will teach them to manage time and resources effectively, work harmoniously with others in order to achieve a common purpose, improve their creativity and lateral thinking skills, and lead to the development of greater self-confidence. Experience gained in a number of third world countries has shown that Technology can be used to motivate children to improve their language, mathematical and scientific knowledge and skills.

There is currently no formal technology curriculum in South African schools. Several pilot projects have been initiated, specifically by the Natal Education Department, by a number of the independent schools and by some non-governmental organizations. At this stage much of the work done in the pilot projects has been carried out in schools which are well staffed and reasonably well equipped. The problems facing the curriculum designers are enormous. Any effective national or regional curriculum will have to meet the needs of high quality schools as well as those schools in the Black townships and rural areas which are often poorly staffed and are almost totally lacking in facilities. There will have to be a strong emphasis on technology as a process which leads to the acquisition of appropriate knowledge, skills and attitudes. Projects will have to be developed which draw on resources and materials which are cheap and readily available.

Given the pressures facing the government, and the need to restructure the present education system, it is important that all available resources should be used to optimum effect. There is clearly a role for non-governmental organizations in the introduction of technology to the school curriculum.

4 The ORT-STEP Schools Consultancy Service

The ORT-STEP Institute (see Dr. Eisenberg's paper "Development and Implementation of a Model for technology education in South Africa - The ORT-STEP Experience") was opened in May 1993 with the primary mission of training and retraining teachers to teach technology.

From the outset it was realized that the training of Technology teachers was necessary but not sufficient. Given the fact that there has never been a Technology component in the South African curriculum, and that many of the teachers have little practical experience, it was clear that the theoretical and practical experience gained on the course would have to be supplemented by providing support in the school situation. One of the biggest problems facing teachers, who have a limited technological background, is a lack of confidence. They require ongoing encouragement, support and reinforcement. All too often teachers are sent out from teacher training institutions and left to sink or swim. This is particularly true in a third world situation where peer support is usually non-existent.

The ORT-STEP Schools Consultancy Service was established in February 1994 to support the teacher training programme. The consultancy service provides a support base for practising teachers who have trained at the institute and enables them to call on the institute for help and advice if they run into difficulties with courseware, software or hardware.

ORT-STEP provides a three part support system:

– qualified tutors from the institute pay routine visits to teachers in their schools.
– teachers are able to contact the institute, at any time, to discuss problems they are experiencing with the material they are using in their courses.
– a technical service team is on call to deal with any problems associated with the software, hardware and technology equipment.

In addition, the consultancy service provides support and advice to those schools which have shown an interest in developing technology programmes. The service advises the schools about the steps to be taken in establishing science and technology laboratories and in planning curricula.

One of the main functions of the Schools Consultancy Service is to establish laboratories in selected satellite schools.

5 The ORT-STEP Satellite Schools Project

The ORT-STEP Institute seeks funding from the private sector for the establishment of science and technology laboratories in schools which serve those sections of the community which were disadvantaged by apartheid. The programme started in January 1994. Funding was obtained from the Anglo American and De Beers Chairman's Fund to establish laboratories in six satellite schools. It is anticipated that further funding from various companies and educational trusts will be available in 1995.

5.1 Criteria for the Selection of Satellite Schools

The Role of the School Principal. The attitude of the Principal towards technology education is a critical factor in the success of the project and the selection of a satellite school. If the introduction of technology to the school curriculum is to be successful, the Principal must have an understanding of the underlying philosophy and a clear vision for the role of technology in the school's educational programme. Given the sea of misunderstanding which exists in South Africa, most attempts to introduce technology are initially met with resistance by those who equate it with narrow technical training. The Principal has to sell the concept of technology education to his Governing Body, staff and parents. It is particularly important that the concept of technology education is understood and accepted by members of the school's staff. Failure to create the right climate will almost certainly lead to the collapse of any programme to introduce Technology. Some vigorous Technology programmes, established by totally committed and enthusiastic Principals, have collapsed when their successors have been lukewarm about the subject. technology education in South Africa is in its infancy, is not yet an official subject in the school curriculum, and requires whole-hearted and positive support from the Principal of the school.

In addition to the above, the introduction of technology has implications with regard to timetabling, staffing and the provision of suitable space which can only be resolved with the Principal's active co-operation and support. Technology must be seen to be an important and integral part of the school's curriculum.

The Availability of Technology Teachers. Prior to the establishment of the ORT-STEP Institute there were no facilities available for the training of technology teachers as opposed to teachers of technical subjects, in South Africa. In order to ensure the success of the programme all satellite schools are required to have at least two teachers on the ORT Higher Education Diploma Course in Technology, Science and Mathematics Education.

The Provision of Suitable Facilities for Technology Education. The budget available for the establishment of technology laboratories does not permit the institute to provide new buildings. The satellite schools are required to provide a suitable room. The institute provides funding for the furnishing of the laboratory, the establishment of the necessary services, such as power points, and the provision of appropriate equipment. In addition to the laboratory, or laboratories, the provision of adequate and carefully planned storage is an essential, and often neglected, element in the implementation of a successful Technology programme. All work is carried out in conjunction with the school authorities and is supervised by the Schools Consultancy Service.

The equipment is supplied on a phased basis and is determined by the teachers' progress on the course and their ability to use the equipment effectively.

Academic and Attendance Records. Many schools in the Black townships have been adversely affected by boycotts, often accompanied by intimidation and violence, the destruction of school property and negative attitudes towards education. Many years of apartheid education, the politicization of the youth, together with recent unrest in schools coupled with political slogans such as 'Liberation Before Education' and 'Pass One Pass All' have done little to establish a culture of learning. In selecting the satellite schools, care has been taken to identify those schools which have managed to maintain sound academic and behavioural standards in spite of the difficult circumstances in which they operate.

Community Involvement. All satellite schools are required to share the facilities in their technology laboratories with neighbouring schools within the constraints imposed by the host school's needs. It is envisaged that, as the programme develops, the facilities will be used for the training of teachers. This is seen to be particularly important if there is to be the necessary exponential growth in the training of Technology teachers. The satellite schools are also expected to make the facilities available to members of the community, which the school serves, by developing appropriate adult education programmes.

6 Conclusion

Present attitudes towards technology education need to be improved if South Africa is to be competitive in the latter part of the 20th century and in the 21st century. The training of teachers lies at the heart of any programme to introduce technology education to the school curriculum. Given the lack of experience in the country it is essential that all technology teachers are be supported in the classroom situation. The scarcity of resources makes it particularly important that they should be used as effectively and efficiently as possible.

The government is currently preoccupied with the need to restructure education. This, together with the current lack of expertise in technology education, makes it appropriate for non-governmental organizations having the necessary background and experience to become involved in the training of Technology teachers and the provision of consultancy and support services.

The current initiatives are pilot projects and will have to be subjected to ongoing and critical assessment in order to ensure that they meet the educational needs of the communities they are designed to serve.

References

1. Malherbe, E. G.: Education in South Africa. Volume 2 (1923-1975). Juta 1977

Conference Paper Presentations: Summary

C. D. Mockford, United Kingdom, T. Ginner, Sweden and P. van Schalkwyk, South Africa

A wide variety of interesting papers have been presented at this WOCATE conference, all relating to different aspects of school technology education. They have covered a number of topics vital to the continued and successful development of this powerful area of education within the world context.

In this short report from the rapporteurs of the three parallel paper presentations it is difficult to give each author appropriate consideration. We have therefore endeavoured to suggest key topics and areas of interest that will lead readers to a more substantial consideration of individual papers.

The sub-titles used in this report reflect the special areas addressed by paper presenters during the conference. The names of paper authors who contributed to these areas of debate concerning technology education are listed at the end of the section. Readers can locate the full paper through this connection.

Keynote presentations are not reported here: rather, our commentary relates to the presentations of the general conference delegates.

1 School Technology Education

1.1 Technology Education at Policy Level

Judging from the papers presented there are quite a few similarities and parallels in the development of technology education in the countries represented at this conference. Even if the emphasis changes from one country to another there seems to be considerable agreement concerning the aims and objectives of technology education. A significant number of countries share several important goals; in particular, a desire for a broader perspective on technology and technology education, and recognising, for instance, the necessity to change aims, content and methods of teaching the subject in order to make the learning area more interesting.

The fact that the subject of technology education is embedded in cultural and socio-economical contexts that vary from country to country gives educators and politicians different points of departure. It is also clear that the technology education community has to recognise that stakeholders with sometimes contradictory interests do not always have the same power to influence the development of the school curriculum. The risk of adopting a narrow perspective of the subject in a desire to achieve what can be considered to be short term economic aims is a danger that was explored in paper presentations.

In the papers from some east European countries another common problem was presented. This concerned the fact that the general public, politicians and even people within the school system tend to connect the subject of school technology to the old political system. This can create difficulties for those educators striving to implement a 'new' programme of school technology.

As a world community, even if we have a lot in common there are, and must be, many differences. That is especially true when it comes to issues of organisation such as whether to make the subject compulsory or not, and whether to introduce technology in the school system at primary, secondary or later phases.

To develop the debate, it would be interesting and potentially helpful to complete a closely focused, structured comparison between the curricula in different countries. This would include three key areas for consideration: from where does each country embark (traditions, school systems etc.); how do they define technology and school technology; what are the aims and targets for the subject.

(D. Ferguson, New Zealand; T. Ginner, Sweden; G. Höpken, Germany; J. Hudec, Slovak Republic; L. Radics, Hungary; M. Vala, Czech Republic)

1.2 School-based Technology Education

A number of papers considered technology education from a more theoretical stand point. Work was presented that stressed the importance of integration in technology education; that is the integration of methods, disciplines and approaches, mind and hand, problems and phenomena. This theme relates closely to the work that is highlighted later in this report under the heading of methods, didactics and equipment, where the importance of the integrative role that technology education can perform is discussed further.

The relationship between theoretical knowledge and practical abilities continues to engage educationalists in the area of technology education. Whether to teach broad but relatively superficial theoretical knowledge or to adopt strategies that present learners with the opportunity to develop narrow but deep knowledge continues to be a topic for debate. This is reflected in the consideration of whether we are teaching for technological awareness or capability.

Across countries of the world the language and terminology associated with technology education often carry different meanings, an example being the difference between the words designer, technologist and engineer. In some languages it would appear that there is a need for a technology education vocabulary which, taken to its logical conclusion, would suggest a need for a world technology education vocabulary of discourse, understood by all, facilitating effective communication and understanding amongst educationalists working in this emergent field. This, however, may be more utopian than realistic.

Issues relating to motivation in technology education were discussed, especially the relationship between general school performance and achievement motivation. Work in the field of psychology, dealing with cognitive mapping, led to the development of a framework by which technology education could be analysed. This framework provides data that are useful in determining instructional design and learning styles which may be more effective in the field of technology education.

In the context of the range of subjects that constitute the school curriculum, technology is a relatively new curriculum area. Its associated research base needs to be developed through further, substantive work in order to provide firm evidence to support the claims that are made for its importance. In this regard, looking at philosophical aspects and developing this strand of technology education is also important for the future of the subject.

(J. Dubovská, Slovak Republic; M. Duris, Slovak Republic; M. Flesková, Slovak Republic; V. Georgieva, Bulgaria; B. Hill, Germany; F. Mosna, Czech Republic; I. Natali, Italy; S. Pillai, India; J. Stoffa, Slovak Republic; P. van Schalkwyk, South Africa)

1.3 Methods, Didactics and Equipment

In the field of methods and didactics several important issues were discussed. One - eternal it seems - concerned integration. In the field of technology education the concept of integration is complicated. The papers that were presented and discussed highlighted several different perspectives: the integration of different materials; the relationship between technology and other subjects; the integrative connections between content and process.

It is often far from clear what we mean when we talk about integrative learning. It can be a term used to describe a teacher's presentation that is integrated in one way or another, or the integration of different materials within a subject, or the integration of different subjects into one curriculum area. The range of topics and interpretations can be varied and wide ranging. Throughout this debate the crucial point is the integration of knowledge that the student finally forms or shapes in her or his own mind. Whilst teachers can support this process in different ways they can never control it.

One of the most important questions in education, not least in connection with school technology, is about creativity. Teacher training does not always encourage and actively promote creativity and the means by which teachers are able to help their pupils to develop this important feature of human behaviour. Creativity was discussed explicitly and implicitly by a number of paper presenters, and it is clear that most countries recognise that there is a need to engender a creative attitude in pupils. Technology education has a key role to play in this process.

What physical materials and resources to use in technology education is, of course, a question closely linked to the issues discussed above. The presentations gave two totally different but interesting alternatives. The paper concerning the use of the LEGO system showed how this useful resource can be used effectively in technology education. An important question in connection with all the kit-based teaching resources is which skills the pupils really develop and how this is achieved. The educational and methodological topics are of central interest, but a more practical, pragmatic issue is that of cost.

The other presentation on equipment exposed a quite different approach, involving modern information technology that can be used to give students and teachers access to vast amounts of information. In addition, it was suggested that the use of information technology can give pupils an opportunity to connect to real problem situations, experiencing a more informed and complex problem solving context. The restricted budgets that schools normally work under are quite often a crucial obstacle when there is a desire to explore the use and potential learning benefits of these powerful information technology based resources.

(J. Barnes, USA; J. Benusková, Slovak Republic; M. Kozuchová, Slovak Republic; J. Polák, Slovak Republic; M. Toman, Slovak Republic; K. Uzdzicki, Poland)

1.4 School Technology and Life

Perhaps more than many other subjects in the school curriculum, technology has the potential to deal with topics and problems close to real life. This can be the case through from elementary to tertiary levels of education. There are two significant issues related to this theme of the conference. First, to make technology education interesting and successful is it a necessity to deal with the world of work? Second, if the agencies influencing education demand the inclusion of more work-related experiences in the school education is technology the most appropriate subject to deliver this? The papers presented discussed both issues.

From listening to the papers presented under this title an interesting issue emerged. While some countries in the western world want their schools to be more closely linked to industry and to create economic understanding among pupils, some examples from eastern Europe show strong ambition to break any very close connection between industry, vocation and school education. It will be interesting to see how these two tendencies develop.

(C. Benson, United Kingdom; M. Jakowicka, Poland; S. Kiss, Hungary)

2 Teacher Education in Technology Education

Teacher education and training was identified by a number of speakers as a key element in the process of change, contributing to the delivery of a high quality programme of education in schools. Concern about the standard of entrants to training programmes and the retention of these trained but relatively small number of professionals was a common theme in many of the papers presented. It was encouraging to note that once engaged in a teacher training course, students' motivations were usually very high, although the carry forward to employment was not always what was expected, with students regularly seeking employment in areas of professional activity other than teaching.

The content of teacher education courses also received consideration in papers presented during the conference. The balance between school-based activities, didactics and subject knowledge acquisition always causes considerable debate. This seems to be especially so in technology education where a broad range of topics can be included in the curricula of schools, leading to questions of coverage through training. Whilst a broad coverage is required to give new teachers an awareness of many areas of technology, it is also essential to ensure that teachers emerge who are competent and confident with their subject. This may require teacher education courses to provide opportunities for special domains of knowledge and skills to be developed during training.

The action of communicating different models of teacher education that are implemented in countries across the world gave other educators the opportunity to reflect upon these programmes and, where appropriate, to adopt different features of them. However, it was realised that this translation of ideas must take place in the context of the culture of that country, maintaining tradition, but developing a new framework for technology education.

From consideration of the papers presented that were linked to teacher education there emerged a number of central issues. In particular, the retention rates for technology teachers, the content of teacher education courses and the effects that changes in teacher education programmes have on the curricula of schools once the students move into professional practice warrant further consideration.

(P. Ferko, Slovak Republic; B. Rychlikova, Czech Republic; T. Ginner, Sweden)

3 Technology Education in the University Sector

Within the university sector, one focus of attention was the development of new technology education programmes that sought to provide the next generation of engineers and technologists. In addition, due regard was given to the need to educate all university students about technology and its potential for change, including its impact on both people and the environment. Technology education and technological thinking within all university courses can be seen to be a vital part of the education of all professional practitioners no matter in which field they may specialise.

In particular, discussion centred on how to incorporate changes in technology into syllabuses that are already overcrowded. These papers focused on courses that were concerned with automated production and digital electronics.

A need was also identified to move towards different styles of teaching and learning within the university sector of technology education. It was suggested that urgent consideration should be given to how students might be supported in developing more creative approaches to the application of technology and more confident attitudes when dealing with technology. This will be particularly important if countries are to have the future technologists who will fashion the products of the coming years.

Information technology, it was agreed, has a vital role to play in developing these styles of learning and teaching, whilst at the same time widening access to technological education across different age groups and phases of employment. Developments in the use of multimedia, especially CDi were suggested as a potential mechanism for delivering new programmes of technology education in a form that can give wider access by comparison to traditional school and university based courses.

The support that the EC COMETT programme has provided for technology transfer and training was highlighted by practitioners involved with this scheme. As co-operation continues to increase and the benefits to both industry and education become apparent, so the programmes will attract a wider set of participants. However, there is a need to identify, validate and communicate the potential and likely actual value of these exchange programmes to both education and industry partners.

Issues for the future are focused towards two major stands: the development of new technological curricula to meet the challenge of changing technologies; the use of new technology to provide exciting learning opportunities and wider access to technology education.

(F. March and M. Künne, Germany; C. Mockford, United Kingdom; G. Pál, Hungary; V. Poppeová, Slovak Republic)

4 Technology Education and Industry

The conference engaged a number of delegates in debate concerning the effects of technology education in schools and the relationship to working practices in industry. In particular, the power of technology education to bring about a change in working practices was highlighted. In the climate of change that is evident across the world, but especially in the east European countries, technology education was identified as a special subject that has a valuable role to play. This view was supported by those operating in industries where change is urgently needed. The need for re-training company employees both across the age range and at all levels was stressed if prosperity in a free-market economy is to be achieved and maintained.

Proposals for training were suggested in the context of developing professional competences for those working in industries where new styles of management have been introduced. Looking towards the future, more research needs to be conducted concerning the effects, both positive and negative that result from these changes that are taking place in company training.

(A. Dingová, Slovak Republic; W. E. Theuerkauf and A. Weiner, Germany)

Resolution Adopted by the Conference:
Technological Education for All

In recognition of technology as one of the most powerful forces affecting the quality of life in all nations of the world, and building upon the formal resolutions of the delegates attending the Formal Constitution Ratification Conference in Paris and the Project 2000 + Meeting at the UNESCO Headquarters in July 1993, we, on behalf of the WOCATE members and the 107 delegates, from 22 countries, attending the Banska Conference duly declare that:

- The quality of life afforded by a society is directly and positively related to the extent to which its people understand, effectively use and develop new technologies in correlation with economical, social and ecological balances.

- Technology education develops critical survival skills and insights for citizens in a society dominated by change and, as an essential discipline, technology must be provided at all levels of the educational system, as well as by continuing education programs.

- Technology is a discipline in its own right and as such technology education should become an acknowledged subject for the development of human potential and capabilities beginning with education for all and extending through all levels.

- In order to achieve understanding of technology, a significant insight must be established into the complex relationships found in technological situations, such as, economic, scientific, environmental, and practical ones, etc., and their appropriate equilibrium.

To underline the preceding views, the participants of the WOCATE Conference formally resolve to assume a global responsibility, to develop an appropriate framework to satisfy contemporary demands related to polyvalent technological education on all levels. This framework must seek:

- To empower humans through technology education to make appropriate use of technological advancements.

- The implementation of systematic strategies and programs that address the growing international crisis caused by a critical undersupply of teachers trained for the implementation of technology education programs, at all levels, and as appropriate to the needs of each country,

- The design of educational strategies needs to address elements such as education - industry co-operation, the integration of related subject areas, and the use of learning situations vested in a real world context.

- The development of appropriate educational strategies as well as, the creation of networks for the transfer, the transformation and diffusion of information related to advanced practices in technological education.

- To continue the development of technology education through systematic, world-wide, and collaborative programs of research into the essential elements of, and conditions for, technological literacy.

- The continuation and strengthening of measures within technology education to address issues relating to its equity, accessibility and effectiveness with respect to gender, race and creed.

All parties, agencies, organizations, corporations and ministries are invited to join with us in the international thrust to create systems that develop people adequately prepared to understand, use, develop and control technology in its complex context rather than to be controlled by it.

This declaration was formally ratified by the International Conference on

*"Technology Education, Innovation and Management"
held in Banská Bystrica from 24th to 29th September 1994.*

List of Participants

James L. Barnes, Dr.
Professor
Eastern Michigan University
NASA Research Center
Starkweather Hall
Ypsilanti, MI 48197
USA
phone: +1-313-4 87 24 69

Jozef Belák, Dipl.-Ing.
Ministry of Economy of the
Slovak Republic
Department for Technical Policy
Mierová 19
SK-82715 Bratislava
Slovak Republic
phone: +42-7-23 40 00

Clare Benson
University of Central England
Faculty of Education
Westbourne Road Edgbaston
Birmingham B15 3TW
United Kingdom
phone: +44-21-3 31 61 30

Jolana Benusková, Dr.-Ing.
Matej Bel University
Faculty of Education
Zvolenská Cesta 6
SK-97400 Banská Bystrica
Slovak Republic
phone: +42-88-76 48 45

Dietrich Blandow, Dr.
Professor
WOCATE Executive Secretary
WOCATE Office Erfurt
Schlösserstr. 9
D-99084 Erfurt
Germany
phone: +49-361-5 62 10 82

Ellen Bommersheim
J. W. von Goethe University
Senckenberganlage 15/60054
D-60325 Frankfurt/Main
Germany
phone: +49-69-7 98 33 93

Rosemarie Burow
Association for the Promotion
of Work Oriented Research and
Education
Am Eschbachtal 50
D-60437 Frankfurt/Main
Germany
phone: +49-69-7 98 33 93

Arne Cako, Dr.-Ing.
College of Education
Department of Technology
Education
Hornocermánska 4
SK-94974 Nitra
Slovak Republic
phone: +42-87-52 99 53

Arthur Cotton
ORT-STEP Institute
Schools Consultancy Service
Private Bag X 13
1685 Halfway House
Republic of South Africa
phone: +27-11-6 51 65 36

Wolfgang Deutz, Dipl.-Ing.
IBS
Rathausstraße 69
D-56203 Höhr-Grenzhausen
Germany
phone: +49-2624-9 18 00

Alzbeta Dingová, Dr.
Chirana - Prema a.s.
Factory for Medical Instruments
Nám. Dr. Alberta Schweitzera 194
SK-91601 Stará Turá
Slovak Republic

Dusan Dobrovodsky
Programme Manager
PHARE Coordination Office
Bratislava
Sládkovicova 3
SK-81106 Bratislava
Slovak Republic
phone: +42-7-36 35 98

Rozmarín Dubovská, Dr.
Matej Bel University
Department of Technology
Education
Tajovského ul. c. 40
SK-97549 Banská Bystrica
Slovak Republic
phone: +42-88-3 58 60

Milan Duris, Dr.
Matej Bel University
Department of Technology
Education
Tajovského ul. c. 40
SK-97549 Banská Bystrica
Slovak Republic
phone: +42-88-3 58 60

Michael Dyrenfurth, Dr.
Professor
Technology & Industry Education
University of Missouri-Columbia
105 London Hall
Columbia, MO 65211
USA
phone: +1-314-8 82 278 2

John Eggleston, Dr.
Professor
Warwick University
Department of Education
Westwood
Coventry CV4 7AL
United Kingdom
phone: +44-203-52 41 04

Matzi Eliahu, Dr.
Center for Technological Education
affiliated with Tel-Aviv University
Department of Education
53 Golomb Street
Holon, 58386
Israel
phone: +972-3-5 02 80 27

Don Ferguson
New Zealand Ministry of Education
Department of Policy
P.O. Box 1666
Wellington
New Zealand
phone: +64-4-4 71 60 41

Pavol Ferko, Dr.
Professor
Matej Bel University
Faculty of Natural Science
Tajovského ul. c. 40
SK-97549 Banská Bystrica
Slovak Republic
phone: +42-88-3 45 53

Marta Flesková, Dr.
Matej Bel University
Faculty of Education
Zvolenská Cesta 6
SK-97400 Banská Bystrica
Slovak Republic
phone: +42-88-6 16 31

Diny Flierman
Foundation "Ontdekplek"
Zoetestraat 11
2011 PP Haarlem
The Netherlands
phone: +31-23-31 90 30

Katja Gerlach
Association for the Promotion
of Work Oriented Research and
Education
Am Eschbachtal 50
D-60437 Frankfurt/Main
Germany
phone: +49-69-7 98 33 93

Vanya Georgieva, Dr.
Ass. Professor
South West University
"Neofit Rilski"
Engineering Pedagogical Faculty
A. Velichov Street 66
2700 Blagoevgrad
Bulgaria
phone: +359-73-2 45 68

Thomas Ginner, Dr.
Linköping University
Department of Technology and
Social Change
S-58189 Linköping
Sweden
phone: +46-13-28 10 00

Paul Griffiths
University of Brighton
School of Education
Falmer
Brighton BN1 9PH
United Kingdom
phone: +44-273-643 3 88

Friedhelm Hammerschmidt,
Dr.
Deutsche Gesellschaft für
Technische Zusammenarbeit (GTZ)
Postfach 5180
D-65726 Eschborn
Germany
phone: +49-6196-79 11 67

Bernd Hill, Dr.
College of Education Erfurt
Institute of Technical Science and
Company's Development
Nordhäuser Straße 63
D-99089 Erfurt
Germany
phone: +49-361-7 37 16 55

Gerd Höpken, Dr.
University of Flensburg
Institute for Technology and Its
Didactics
Fruerlunder Straße 37
D-24943 Flensburg
Germany
phone: +49-461-3 89 90

Agnieszka Horsey
The Standing Conference on
Schools' Science and Technology 76
Portland Place
London W1N 4AA
United Kingdom
phone: +44-71-2 78 24 68

Ján Hudec, Dr.
Matej Bel University
Department of Technology
Education
Tajovského ul. c. 40
SK-97549 Banská Bystrica
Slovak Republic
phone: +42-88-3 58 60

Richard Huisinga, Dr.
Professor
Association for the Promotion
of Work Oriented Research and
Education
Am Eschbachtal 50
D-60437 Frankfurt/Main
Germany
phone: +49-69-7 98 33 93

Sylvia Innes
The Standing Conference on
Schools' Science and Technology
Professional Development Unit
76 Portland Place
London WIN 4AA
United Kingdom
phone: +44-71-2 78 24 68

Maria Jakowicka, Dr.
Professor
Pedagogical University
Department of Pedagogy
Al. Wojska Polskiego 69
65325 Zielona Góra
Poland
phone: +48-68-6 35 20/134

Sándor Kiss, Dr.
Kölcsey Református Teachers
Training College
Department of Technology
Péterfia 1-7
H-4026 Debrecen
Hungary
phone: +36-52-41 81 66

Dieter Koch
J. W. von Goethe University
Senckenberganlage 15/60054
D-60325 Frankfurt/Main
Germany
phone: +49-69-7 98 33 93

Frank König
Company Ramona König
Medical Bandages
Hauptstraße 46
D-07422 Milbitz
Germany
phone: +49-36739-2 22 08

Maria Kozuchová, Dr.
Comenius University
Faculty of Education
Moskovská 3
SK-81334 Bratislava
Slovak Republic
phone: +42-7-22 11 24

Zuzana Krajcusková, Ing.
Slovak Technical University
Faculty of Engineering
Ilkovicova 3
SK-81219 Bratislava
Slovak Republic
phone: +42-7-35 11 37

Ivan Kruspan, Ing.
Matej Bel University
Department of Technology
Education
Tajovského ul. c. 40
SK-97549 Banská Bystrica
Slovak Republic
phone: +42-88-3 58 60

Helena Kubátová
Research Institute of Vocational and
Technical Education
Karlovo nám. 17
CZ-12000 Prague 2
Czech Republic
phone: +42-2-29 18 06

Matthias Künne, Dipl.-Ing.
COMETT UETP Saxony
c/o Technical University Dresden
Mommsenstraße 13
D-01139 Dresden
Germany
phone: +49-351-4 63 22 18

Kati Langer, Dr.
WOCATE Office Erfurt
Schlösserstraße 9
D-99084 Erfurt
Germany
phone: +49-361-5 62 10 82

Anton Lavrin, Dr.
Technical University of Kosice
Letná 9
SK-04187 Kosice
Slovak Republic
phone: +42-95-5 35 74

Roman Linczéni, Dipl.-Ing.
Director
Business & Innovation Centre
Bratislava
Nevädzová 5
SK-82101 Bratislava
Slovak Republic
phone: +42-7-23 76 66

Ingrid Lisop, Dr.
Professor
Association for the Promotion
of Work Oriented Research and
Education
Am Eschbachtal 50
D-60437 Frankfurt/Main
Germany
phone: +49-69-7 98 33 93

Manfred Lutherdt, Dr.
Professor
College of Education Erfurt
Institute of Technical Sciences and
Company's Development
Nordhäuser Straße 63
D-99089 Erfurt
Germany
phone: +49-361-7 37 11 14

Frank March, Dr.
COMETT UETP Thuringia
c/o Technical University Ilmenau
Max-Planck-Ring 14/PE 327
D-98684 Ilmenau
Germany
phone: +49-3677-69 25 25

Bernd Mendritzki
J. W. von Goethe University
Senckenberganlage 15/60054
D-60325 Frankfurt/Main
Germany
phone: +49-69-7 98 33 93

Matthias Metzing, Dr.
WOCATE Office Erfurt
Schlösserstraße 9
D-99084 Erfurt
Germany
phone: +49-361-5 62 10 82

Pavol Micúch, Dr.
Skolska sprava
Department of Methodology
Stefánikova 160/2
SK-01701 Povazská Bystrica
Slovak Republic
phone: +42-822-2 37 71

Milan Mikosek, Dipl.-Ing.
University of Ostrava
Faculty of Education
Dvorákova 7
CZ-70103 Ostrava 1
Czech Republic
phone: +42-69-6 22 27 55

Roland Mildes
Association for the Promotion
of Work Oriented Research and
Education
Am Eschbachtal 50
D-60437 Frankfurt/Main
Germany
phone: +49-69-7 98 33 93

Clive D. Mockford
Loughborough University
Department of Design and
Technology
Loughborough
Leicestershire LE11 3TU
United Kingdom
phone: +44-509-22 26 69

Kevin Morgan, Dr.
Professor
Griffith University
School of Technology Education
Nathan, Brisbane, 4111
Australia
phone: +61-7-8 75 58 25

Frantisek Mosna, Dr.
Czech Association of Technology
Education Teachers
c/o Charles University Prague
Faculty of Education
M. D. Rettigové
CZ-11639 Prague
Czech Republic
phone: +42-2-24 91 56 17

Ilia Natali
Professor
International Institute for
Technology Education
c/o CEDE
Villa Falconieri
I-00044 Frascati-Rome
Via Borromini, 5
Italy
phone: +39-6-9425771

Ondrej Nemcok, Dr.
Matej Bel University
Department of Technology
Education
Tajovského ul. c. 40
SK-97549 Banská Bystrica
Slovak Republic
phone: +42-88-3 58 60

Hana Nováková
Charles University Prague
Faculty of Education
M. D. Rettigové 4
CZ-11639 Prague
Czech Republic
phone: +42-2-24 91 56 17

Dusan Ondrejicka
Delegation of the European
Commission
PHARE Coordination Office
Sládkovicova 3
SK-81106 Bratislava
Slovak Republic
phone: +42-7-36 36 20

Gabriella Pál
Budapest Training Technology
Centre
Bródy Sándor U. 14
H-1088 Budapest
Hungary
phone: +36-1-1 18 65 22

Haris Papoutsakis
Ass. Professor
Technical Education Institute of
Heraklion
P.O. Box 140
GR-71110 Heraklion
Greece
phone: +30-81-25 41 25

Ján Pavlovkin, Ing.
Matej Bel University
Department of Technology
Education
Tajovského ul. c. 40
SK-97401 Banská Bystrica
Slovak Republic
phone: +42-88-3 58 60

Pavel Petrovic, Dr.
Research Institute of Vocational and
Technical Education
Karlovo nám. 17
CZ-12000 Prague 2
Czech Republic
phone: +42-2-29 18 06

Swaminatha Pillai, Dr.
Professor
Technical Teachers' Training
Institute
Department of Educational Research
Taramani
600113 Madras,Tamilnadu
India
phone: +91-44-2 35 20 54

Hans-Christian Piossek
Municipal Corporation Erfurt
Fischmarkt 1
D-99084 Erfurt
Germany
phone: +49-361-6 55 10 06

Jozef Polák, Dr.
College of Education Nitra
Ir. A. Hlinku 1
SK-94974 Nitra
Slovak Republic
phone: +42-87-51 10 13

Eva Poláková, Dr.
College of Education Nitra
Ir. A. Hlinku 1
SK-94974 Nitra
Slovak Republic
phone: +42-87-51 10 13

Viera Poppeová, Dr.
University of Transport and
Communications
Department of Measurement and
Automation
Velky Diel
SK-01026 Zilina
Slovak Republic
phone: +42-89-5 28 06

Alfred Prigl, Dr.
College of Zilina
Moyzesova c. 20
SK-01026 Zilina
Slovak Republic
phone: +42-89-65 72 88

Pavel Prokopovic, Dr.
Ass. Professor
Technical University of Kosice
Faculty of Professional Technical
Studies
nám. Mieru 3
SK-08001 Presov
Slovak Republic
phone: +42-91-22 603 5

Lajos Radics, Dr.
Professor
Eszterházy Károly T.
Képzö Föiskola
Szabadság u. 2
H-3300 Eger
Hungary
phone: +36-412-39 91 08

Berta Rychlíková, Dr.
University of Ostrava
Faculty of Education
Dvorákova 7
CZ-70103 Ostrava
Czech Republic
phone: +42-69-6 22 27 55

Tomás Sabol, Dr.
Vice-Chancellor for International Relationship
Technical University of Kosice
Letná 9
SK-04187 Kosice
Slovak Republic
phone: +42-95-5 35 74

Peter Schmid, Dr.
Professor
Eindhoven University of Technology
Faculty of Architecture and Building Sciences
Postbox 513, (Int. Postvak 8)
5600 MB Eindhoven
The Netherlands
phone: +31-40-47 23 73

Burkhard Schweiz
Association for the Promotion of Work Oriented Research and Education
Am Eschbachtal 50
D-60437 Frankfurt/Main
Germany
phone: +49-69-7 98 33 93

Rodney F. Sherwood
Natal Education Department
Research and Development Unit
P/Bag 9044
3200 Pietermaritzburg
Republic of South Africa
phone: +27-331-94 23 51

George Shield
University of Sunderland
School of Education
Hammerton Hall
Gray Road
Sunderland SR2 8JB
United Kingdom
phone: +44-91-5 15 23 93

Jozef Siska
College of Education Nitra
Faculty of Natural Sciences
Ir. A. Hlinku 1
SK-94974 Nitra
Slovak Republic
phone: +42-87-51 10 13

Ivan Spudil, Ing.
Slovak Technical University
Faculty of Electroengineering
Ilkovicova 3
SK-81219 Bratislava
Slovak Republic
phone: +42-7-35 11 19

Subject Index

activities
    educational 210,
analysis
    competitive situation 169,
    market 169,
    SWOT 169,
attitudes
    teacher and student 216,
bionics 101,
college 273,
COMETT 266,
company 210,
competences
    teacher 201,
competition 169,
computer 183,
concept
    course 125,
control
    computer 125,
co-operation
    university-enterprises 266,
creativity 75, 196,
culture
    technological 143,
curriculum 15, 32, 51, 82
    primary 224,
    technology 6, 37
design 147, 224,
    instructional 107,
development
    curriculum 55, 233, 281,
    human resource 75,
dimension
    psychological 107,
education
    economy 241,
    higher 201, 273,
    primary 281,
    reforms in 196, 259,
    secondary 281,
    system of 245,
    structure of 6,
    technology 6, 15, 32, 37, 55, 101, 155, 158, 162, 183, 192, 216, 224, 245, 250, 259, 281, 291,
    vocational 91,
electronics
    digital 147,
elements
    interweaving and coinciding 143,
enlightenment
    technological 137,
gymnasium 15,
industry 216,
industry link projects 224,
informatics 183,
innovation 75, 82, 169, 183, 192,
    educational 196,
internet 82,
key qualification 91,
knowledge
    technological 147,
laboratory
    teaching 158,
law
    education 51,
literacy
    technological 75, 259,
management 137, 210,
mapping
    cognitive 107,
methodology 158,
    innovation 196,
methods
    learning 250,
    problem solving 101,
    teaching 250,
MHPO 162,
motivation 91,
objectives
    curricular 125,
opportunities-threats 169,

ORT-STEP 250,
   Institute 291,
   satellite schools 291,
   School Consultancy Service 291,
partnership
   school/university 201,
personality 117,
place
   discovery 155,
plant
   teaching 125,
preferences 107,
programme
   after school 155,
   EEC 266,
   teaching 147,
project
   NASA 82,
psychical strain 117,
psychohygiene 117,
re-employment 91,
reforms
   curriculum 6,
research
   methodology 55,
school
   comprehensive 245,
   middle 82,
   primary 241,
   secondary 192,
school/industry links 233,
SCSST 233,

studies
   technical 51,
strategy
   combining 169,
   innovative teaching 147,
   transforming or neutralising 169,
structure
   didactical 143,
   knowledge 107,
   networked factory 125,
subject
   curriculum 241,
   implementation 281,
   model 32,
support
   design decision 162,
system
   school 37,
teacher training 233,
   initial (ITT) 201,
teaching 216,
technology 55, 137, 147, 162, 201,
   general 15,
thinking
   development 101,
   technical 259,
training
   teacher 250,
transfer
   technology 266,
understanding
   economic and industrial 224,

Ján Stoffa, Dr.-Ing.
Professor
College of Education Nitra
Faculty of Education
Hornocermánska 4
SK-94974 Nitra
Slovak Republic
phone: +42-87-52 99 53

Walter E. Theuerkauf, Dr.
Professor
University of Hildesheim
Institute of Applied Electrical Engineering and TechnologyEducation
Kreuzstraße 8
D-31134 Hildesheim
Germany
phone: +49-5121-30 44 47

Milan Toman, Dr.
Matej Bel University
Homeland Study Department
Zvolenská Cesta 6
SK-97401 Banská Bystrica
Slovak Republic
phone: +42-88-76 48 46

Doris Traberth
WOCATE Office Erfurt
Schlösserstraße 9
D-99084 Erfurt
Germany
phone: +49-361-5 62 10 82

Kazimierz Uzdzicki, Dr.
Professor
Pedagogical University Zielona Gora
Department of Technology Education
Al. Wojska Polskiego 69
65325 Zielona Góra
Poland
phone: +48-68-6 35 20

Miroslav Vala, Dr.
University of Ostrava
Faculty of Education
Dvorákova 7
CZ-70103 Ostrava
Czech Republic
phone: +42-69-6 22 27 55

Harry Valkenier
Foundation "Ontdekplek"
Highschool Alkemaar
Zoetestraat 11
2011 PP Haarlem
The Netherlands
phone: +31-23-31 90 30

Vladimir Vanek, Dr.
University of Ostrava
Faculty of Education
Dvorákova 7
CZ-70103 Ostrava 1
Czech Republic
phone: +42-69-6 22 27 55

Hana Vanková, Dr.
University of Ostrava
Faculty of Education
Dvorákova 7
CZ-70103 Ostrava 1
Czech Republic
phone: +42-69-6 22 27 55

Peet van Schalkwyk
Professor
Potchefstroom University
Department of Metallurgical Engineering
P/Bag X 6001
2520 Potchefstroom
Republic of South Africa
phone: +27-148-99 16 57

Detlef Wahl, Dr.
WOCATE Office Erfurt
Schlösserstraße 9
D-99084 Erfurt
Germany
phone: +49-361-5 62 10 82

Andreas Weiner
University of Hannover
Institute for Ergonomics and
Didactics of Mechanical Engineering
Im Moore 11 A
D-30167 Hannover
Germany
phone: +49-511-7 62 48 45

Horst Wengel
WOCATE Office Erfurt
Schlösserstraße 9
D-99084 Erfurt
Germany
phone: +49-361-5 62 10 82

Jutta Wilking
Association for the Promotion
of Work Oriented Research and
Education
Am Eschbachtal 50
D-60437 Frankfurt/Main
Germany
phone: +49-69-7 98 33 93

Frantisek Zeleny, Dr.
Ass. Professor
Technical University Zvolen
Faculty of Wood Technology
Woodworking Machine and
Equipment
T. G. Masaryka 24
SK-96053 Zvolen
Slovak Republic
phone: +42-855-2 14 49

Frantisek Zidek, Ing.
Slovak Technical University
Faculty of Electroengineering
Ilkovicova 3
SK-81219 Bratislava
Slovak Republic
phone: +42-7-35 11 74

Volker Zimmermann, Dr.
Association for the Promotion
of Work Oriented Research and
Education
Am Eschbachtal 50
D-60437 Frankfurt/Main
Germany
phone: +49-69-7 98 33 93